SANTA ANA PUBLIC LIBRARY

D0949701

YOUR ATOMIC SELF

ALSO BY CURT STAGER

Deep Future

Field Notes from the Northern Forest

YOUR ATOMIC SELF

THE INVISIBLE ELEMENTS THAT CONNECT YOU
TO EVERYTHING ELSE IN THE UNIVERSE

· ·

CURT STAGER

Thomas Dunne Books ✺ *St. Martin's Press*

New York

THOMAS DUNNE BOOKS.
An imprint of St. Martin's Press.

YOUR ATOMIC SELF. Copyright © 2014 by Curt Stager. All rights reserved. Printed in the United States of America. For information, address St. Martin's Press, 175 Fifth Avenue, New York, N.Y. 10010.

www.thomasdunnebooks.com
www.stmartins.com

Library of Congress Cataloging-in-Publication Data

Stager, Curt.
    Your atomic self: the invisible elements that connect you to everything else in the universe/by Curt Stager.—First edition.
        pages   cm
    Includes bibliographical references and index.
    ISBN 978-1-250-01884-7 (hardcover)
    ISBN 978-1-250-01885-4 (e-book)
    1. Matter—Constitution—Popular works.   2. Human body—Composition—Popular works.   3. Atoms—Popular works.   I. Title.

QC173.S625 2014
530—dc23

2014022128

St. Martin's Press books may be purchased for educational, business, or promotional use. For information on bulk purchases, please contact Macmillan Corporate and Premium Sales Department at 1-800-221-7945, extension 5442, or write specialmarkets@macmillan.com.

First Edition: October 2014

10 9 8 7 6 5 4 3 2 1

Albert Einstein at the Knollwood boathouse on Lower Saranac Lake, NY. This photo, published for the first time, was taken with a Brownie camera during the summer of 1936 by Knollwood resident David Billikopf, then age ten.
Courtesy of David Billikopf

To the Albert in all of us

Contents

Acknowledgments *ix*

Prologue Your Atomic Self *1*

1. Fires of Life *11*

2. The Dance of the Atoms *36*

3. Blood Iron *60*

4. Carbon Chains *86*

5. Tears from the Earth *110*

6. Life, Death, and Bread from the Air *132*

7. Bones and Stones *158*

8. Limits to Growth *182*

9. Fleeting Flesh *206*

Epilogue Einstein's Adirondacks *233*

Notes *247*

References *253*

Index *293*

Acknowledgments

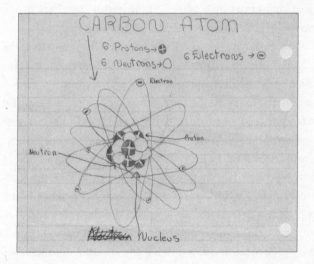

My interest in atoms dates back to my childhood, as the scan of a sixth-grade homework assignment attests. Thanks go to my sister, Leslie, for saving that paper from a trash pile and to the teachers who encouraged my enthusiasm for the sciences in those early days, as well as to all educators who likewise open doors for others.

Thanks to David Schindler for providing the photo of Lake 226, to Suketu Bhavsar, Craig Bohren, Neil Comins, Jeff Couture, Tim Garrett, Charlie Jui, Ralph Keeling, Luiz Martinelli, and Tom Reimchen for informing and fact-checking the text, to Lee Ann Sporn for setting me straight on the inspiration of oxygen, and to Australian radio host Richard Fidler for asking me to follow in words the carbon atoms in a puff of human breath as they travel throughout the world. Thanks also to Paddy and Mike Root for introducing me to Einstein's house, to David Billikopf along with Amy Catania and Mary Hotaling of Historic Saranac Lake, Peter Benson and Michele Tucker of the Saranac Lake Free Library, and Peter Crowley of the *Adirondack Daily Enterprise* for photos and research assistance regarding Einstein's visits to the Adirondacks, to Patty McDonough for training (and maintenance) in the science of hair, and to Hans Bänziger for having the presence of mind to snap the unforgettable photo of his encounter with a tear-drinking moth as well as for letting me share it with you. Kary Johnson, Lee Ann Sporn, Craig Milewski, Laura Rozell, Asha and Jay and Devora Stager, and Bill and Susan Sweeney read the manuscript and gently passed along rich crops of suggestions that have made me look like a better writer than I actually am.

I am grateful to my to-die-for agent and champion, Sandy Dijkstra, along with her associates at the Sandra Dijkstra Literary Agency, for their support and guidance, as well as to my superb editor, Peter Joseph, and the rest of the crew at Thomas Dunne Books and St. Martin's Press who expertly shepherded me and this book through many twists and turns. David Verardo of the National Science Foundation and my "Natural Selections" colleagues, Martha Foley and Joel Hurd, at North Country Public Radio have done much to make me a better science communicator. So, too, have my students and fellow faculty members at Paul Smith's College during the last two and a half decades. These people in addition to the Paul Smith's staff, trustees, and a long list of others have done the North Country a great service by helping to make this remarkable little college on a lake in the mountains an ever-more-vibrant learning environment and a fantastic community to belong to. And as always, my wonderful wife, Kary, has been my most insightful, helpful, and inspiring companion in every aspect of this project as well as in the rest of my life. She takes great photos, too, and has kindly let me use some of them here.

Thank you to Albert Einstein and other scientists both famous and unknown, whose explorations have helped to make this one of the most interesting and transformational periods of human history. And special thanks to all scientists who realize that effective communication with the public is an important skill to be developed and respected, and who consider it an obligation as well as a passion to pursue it.

Finally, here's to those ancestral stars whose relatively short lives and violent deaths made us—and our atoms—what we are.

YOUR ATOMIC SELF

Prologue: Your Atomic Self

Unknowingly, we plough the dust of stars, blown around us by the wind, and drink the universe in a glass of rain.
—Ihab Hassan

One thing I have learned in a long life: that all our science, measured against reality, is primitive and childlike—and yet is the most precious thing we have.
—Albert Einstein

What could be more interesting than the story of your life? Well, good news—that's what this book is about.

Although the text often refers to atoms, they are merely the supporting cast. You are the main character with whom they will interact as you go through the routines, the ecstatic successes, and the profound tragedies of daily life. My role will be to try to show how those same atoms connect you to some of the most amazing things in the universe.

What do atoms have to do with you? Everything. They were present and intimately involved when you and everyone you have ever loved—or hated—did everything that you and they have ever done. Every scent you've ever savored, every sight you've ever seen, every song you've ever enjoyed, every cry or sigh that ever passed your lips sprang from atoms at work within the atmosphere and the darkest recesses of your body. When you eat, the bodies of other living things become part of you. If you cut yourself, the wreckage of dying stars runs out in a stream of ancient atoms that triggered some of the most violent explosions in the cosmos. When you flush your wastes, you scatter the atomic echoes of lightning bolts and volcanoes into a global cycle that may some day return them to you, as unpleasant as that may sound. And whenever you grin, the sparkling of your teeth conceals the dim afterglow of nuclear fallout from Cold War bomb tests over the Pacific.

You are not only made of atoms; you *are* atoms, and this book, in essence, is an atomic field guide to yourself. All you need in order to interpret your life in primal elemental terms is access to some of the latest scientific information, some new ways to reconsider your world in light of it, and an active imagination. In doing so you will begin to experience a revolution in self-awareness that is playing out on a larger scale around the world.

Long after leaving the Neolithic cultures in which we learned to make crude implements from rocks and minerals, we are entering what might be called a "neo-Neolithic" stage in which the exquisitely crafted silicon chips of computers and the polished glass and metal of powerful microscopes and telescopes enhance our lives and inform our senses. With those new tools to help us build upon knowledge left to us by our forebears, we can replace the early Greek concept of four primary elements of creation—air, water, earth, and fire—with a richer and more useful worldview that arranges more than a hundred kinds of atoms into an orderly periodic table of elements. It shows that the first three items on the traditional list are not fundamental elements but compounds, and that fire is more of a process than an indivisible substance. Such a perspective also helps to reveal with scientific rigor the hidden connections that physically link our bodies as well as our very thoughts and feelings to the atoms of the earth. And in this age of intense technological, cultural, and environmental change, knowing just how intimately bound we are to this planet and to one another has become crucially important. Today more than ever, the lessons of science are key to the well-being of billions of people and the ecosystems that sustain them.

Much still awaits discovery, of course, and future research will surely revise much of what we now believe to be true, perhaps including some of the information in this book. Nobody knows everything, and even geniuses make mistakes. Isaac Newton didn't know that matter and energy are interchangeable, Albert Einstein resisted quantum theory, and the physicist Ernest Rutherford, who derided all scientific fields other than his own as mere "stamp collecting," didn't believe that nuclear weapons or power were feasible. Even the meaning of the term "atom" itself changed during the last century or so. More than two thousand years ago, Greek philosopher-scientists deduced that matter consists of tiny fundamental units and called them "a-toms" or "indivisibles." In the strictest sense the things that we now call atoms are unworthy of the name because physicists can split them into smaller pieces. The Large Hadron Collider in Switzerland, for example, smashes subatomic particles into even smaller muons, gluons, and leptons. Therefore, if true a-toms exist, we have not yet found them.

But it would also be incorrect to claim, as some people do, that all scientific facts are too ephemeral to be trustworthy guides to reality. Yes, the frontiers of knowledge are dynamic, but the trailblazers of science leave many reliable paths for you to tread. You can trust, for example, that atoms really do exist. They have distinctive properties and interact in universally predictable ways. Water truly is a combination of hydrogen and oxygen, and so is most of your body. And, yes, the same elements that fill a bead of sweat on your brow can also be found in the majestic tail of a comet, in the bones of the planet beneath your feet, and in every other living thing on Earth. This book invites you to travel some of those trusty paths to remarkable new perspectives on yourself and your world that may be surprising but that are nonetheless demonstrably true and potentially transformative.

Let me introduce you now to one of your oxygen atoms. It won't look like much, basically just a bubble with a speck of matter floating at its center, and from the outside it resembles a featureless Ping-Pong ball. But traits as simple as the number of particles in that central speck or the structure of the outer shell make the properties of such elements as wildly different from one another as the living species they help to produce.

If you are an average-size adult, you carry nearly two thousand trillion trillion oxygen atoms inside you, more numerous than the leaves in every forest on Earth. Imagine singling out one of them from the flesh of your thumb and zooming in for a closer look.

Like all atoms, this one is mostly empty. If you could magnify the jiggling cluster of eight positively charged protons and eight neutral neutrons at its core to the size of a raspberry (which it would then resemble), the negatively charged electrons that encircle it would orbit the berry from about two hundred yards away, roughly twice the length of an American football field. The sphere encompassed by the electrons would contain millions of cubic yards of nothingness, many times the volume of MetLife Stadium in East Rutherford, New Jersey. This is what scientists refer to when they say that the substance of your body is mostly empty, an airless vacuum much like the depths of intergalactic space.

No matter how heavy you may feel you are, the emptiness of your atoms means that you are more like a porous froth of atomic Styrofoam than the relatively solid mass you seem to be. And you are lucky that this is so, because if it were not, the ground beneath you might not support your weight. The tiny central nucleus (Greek for "kernel" or "seed") is so dense that if your body could be packed with deflated atoms rather than the mostly empty ones that comprise you now, the tip of one of your pinkies would weigh close to a billion tons.

Eight electrons circle the nucleus of your chosen oxygen atom. These are not miniature planets limited to a firm orbital plane but stranger, more freely wandering wraiths that act simultaneously as waves and particles, and they swarm around the nucleus at such tremendous speeds that they blur into nested, cloudlike shells. If you were to boost the energy states of electrons such as these, you could make them jump to more distant orbits and make an atom's surface appear to swell, and physicists can also blast electrons away from inner shells with X-ray lasers in order to produce "hollow" atoms.

But most important from our perspective, some of these electrons can be shared with other atoms, forming the covalent chemical bonds that help to hold your body together. Nudge two oxygen atoms close enough to each other for their outer shells to mesh, and some of their electrons may begin to run loops around both nuclei. Such electron-sharing can yield hundreds of thousands of atomic combinations from muscle filaments and membranes to hormones and hairs.

When two or more atoms hook up in this manner, they form a molecule. The term derives from the diminutive version of the Latin word *moles*, which itself refers not to small furry beasts but to a "pile of stuff." A molecule is therefore simply a *small* pile of stuff. The oxygen atom that is now on display in your imagination is rarely alone but is instead normally part of a molecular team. In your body it is most likely to travel in tandem with another oxygen atom as a molecule of oxygen gas in your bloodstream, or to join two smaller hydrogen atoms to form a water molecule.

The first chapters of this book will help you to get to know both of these molecules and their component atoms better, also revealing some of the ways in which they make you what you are. Many atomic elements can be found in and around you, but fewer than a dozen of the most abundant and biologically critical ones will be prominently featured here. These few will serve to unveil the hidden roles that atoms play in your life and show how they connect you to the rest of the world.

As you read on, you will follow your atoms through wind, waves, fire, and forests to your fingernails. Hydrogen atoms will wriggle into your hair and betray where you live and what you have been drinking. The sodium in your tears will link you to long-dead seas and—strangely enough—to fluttering moths. The carbon in your breath will become cornstalks, then the muscles of a mighty bull, and then the twitching whiskers of a fox. You will find that many of the nitrogen atoms in your muscles once helped to turn the sky blue, phosphorus in your bones helped to turn the waves of an ancient ocean green, calcium in your teeth were mined from rocks by mushrooms, and iron in your blood once destroyed a star. You will also discover that much of what death will eventually do to you is already happening among the atoms of your body at this very moment but that, nonetheless, you and everyone else will always exist somewhere in the fabric of the universe.

Taking on this atomic view of life is much like watching an opera with a playbook at hand. If you were to watch *Don Giovanni* without understanding Italian, for example, the music might be enjoyable but the story couldn't grab you. Having a playbook handy would reveal the details of the story that your uninformed eyes and ears would otherwise miss. Similarly, the better you understand the world in atomic terms, the richer your experiences may become.

One of the great gifts of science is the ongoing revelation of what we most truly are. Although it is easy to imagine that cell phones, supermarkets, and city life have isolated us from nature, it is only our awareness of still being connected to it that sometimes lags behind the burst of new technology and knowledge that has erupted during the last century. We can never really lose touch with atomic nature because, whether we know it or not, we are too much a part of it. The task that we now face is not so much to reconnect physically as to more closely attune our worldviews to the fascinating reality that Earth-orbiting telescopes, atom-probing microscopes, and other complex inventions have only recently uncovered for us.

It can be thrilling to revel in the wonder of being direct relatives of the stars, as when one of Joni Mitchell's songs tells us that "we are stardust" and billion-year-old carbon. Those words from the song "Woodstock" capture the essence of insights that science can now confirm, although some nerds among us might also feel compelled to add that most of our carbon is actually much older than that, and that some of it is a few weeks or months old (more on that later).

The main point here is that such insights are aesthetically and philosophically inspiring but also increasingly valuable for practical reasons in this most recent stage of history, which many scientists are calling the Anthropocene epoch, the age of humans. We have become so numerous, our technology so powerful, and our lives and cultures so globally interconnected that we are now a force of nature on a geological scale. Our carbon emissions alone are sufficient to stop the next ice age, to lift the surfaces of the oceans high enough to flood coastlines, and to drive entire species to extinction. And the contents of our hearts and minds now trigger actions that echo around the world and deep into the future.

In this setting, understanding our atomic connections to the earth becomes less a matter of choice than a necessity, and science is our clearest window on the truth. The biologist and author E. O. Wilson recently put it this way: "Perhaps the time has come to cease calling it the 'environmentalist' view, as though it were a lobbying effort outside the mainstream of human activity, and to start calling it the real-world view." Rather than rely solely on our limited senses as our early ancestors did, we can now use new information about the previously hidden atomic nature of things to better interpret what our senses tell us and, we may hope, to produce more sound and sustainable ways of living as well.

Gifted science communicators from Carl Sagan and Neil deGrasse Tyson to Brian Cox, Brian Greene, and Michio Kaku have done a spectacular job of showing us the subatomic realm and the cosmos, but much of the middle ground between those outer limits of size remains to be interpreted from an atomic perspective. Vivid pictures and documentaries help us to imagine tiny quarks and vast galaxies, but it can be difficult to bring them all down to Earth, so to speak. Readily observable species and cells may appear to be simple by comparison to the marvels that mass spectrometers and space telescopes reveal, but most living things are actually too complex and unpredictable to be described with the precision that physicists might expect from their studies of protons or pulsars. Just imagine trying to predict the exact route that a butterfly will follow through a meadow, to calculate exactly when a particular sprouting seed will open its first blossom, or to list in advance all the thoughts that will pop into your head tomorrow. Modeling the motions of electrons and planets is, by comparison, relative child's play.

It is into this lively jungle of daily experience that the following chapters will lead you, armed not with esoteric formulae but with intuitive, sensory, and practical examples of the connections between atoms, yourself, and the world you live in. But how can we make atoms seem real if, as we so often presume, we can't see them?

It's simple. Just look in the mirror and you will see trillions upon trillions of them staring back at you from the contours of your face. If you can recognize atoms in their teeming multitudes for what they are, then you don't have to view them individually in order to sense their presence any more than you need to analyze every grain of sand in order to enjoy a day at the beach. What you can do instead is to let reputable experts work out the more granular details for you and then use what they find to better inform your life.

Amazingly, however, even individual atoms are now more accessible to us than they used to be, and if you want to look one of them in the eye, guides to some of the latest scientific discoveries can help you to do so. The online site Electron Microscopy—A Journey into Nano-Cosmos, hosted by the Triebenberg Laboratory at the Technical University of Dresden, recently presented a series of photographs that zoom in on microscopic specks of gold. The series closes with a final shot that magnifies the sample more than a million times, showing gold atoms arranged like marbles in orderly ranks and files.

Not only are individual atoms becoming visible to us, they are also audible. An online program posted by the Swedish Royal Institute of Technology and titled "The Radioactive Orchestra" allows you to compose a melody using sounds that represent the distinctive energetic frequencies of atoms undergoing radioactive decay. The host site archives tones that represent energy emissions from dozens of unstable elements from carbon-14 to potassium-40. "Our goal is to inspire," the home page explains. "We hope to achieve increased awareness of the beauty of nature, even at its smallest scale, and hence to stimulate interest in basic science. Equally important to us is the creative, musical dimension. . . . There is a lot in common between science and art. We hope the Radioactive Orchestra will contribute to strengthening the bond between the two."

Writing this book has been a personal exploration for me as well. It isn't easy, even for a scientist such as myself who studies the interactions among species, climates, and the elements of life, to connect the invisible to the tangible when I flop onto the couch after a long day at the office or take a walk in the woods. It can be surprisingly difficult to think of oneself as a lump of inanimate matter as well as a person and, truth be told, I have yet to hear many scientists say that they really feel, in their heart of hearts, that they are made of atoms even though they may be able to discuss that fact in depth on an intellectual level. But I assure you that even a fleeting glimpse of this miraculous truth can change you forever, and for the better.

Although my own interest in atoms dates back to early childhood, I began a more transformative journey into atomic reality during the 1970s. I was a college student at the time, majoring in biology and geology, and I was struggling to reconcile my affection for the natural world with the seemingly sterile rigor that my science courses demanded of me. What I didn't yet realize was that such rigor can be as necessary to tuning oneself in to the majesty of life as tedious daily practice is to an actor or musician.

I don't remember who handed me the scrap of paper that contained an excerpt from *The Effect of Gamma Rays on Man-in-the-Moon Marigolds*, a Pulitzer Prize–winning drama written by the playwright and science teacher Paul Zindel in 1964. But I've never forgotten the effect that this fragment of the script had on me. In it, a high school student tells her sister what a teacher recently said to her. These words, I think it is safe to say, changed my life, and they can still produce a lump in my throat when I read them.

> He told me to look at my hand, for a part of it came from a star that exploded too long ago to imagine . . .
>
> When there was life, perhaps this part of me got lost in a fern that was crushed and covered until it was coal.
>
> And then it was a diamond millions of years later—it must have been a diamond as beautiful as the star from which it had first come . . .
>
> And he said this thing was so small—this part of me was so small it couldn't be seen—but it was there from the beginning of the world.

And he called this bit of me an atom. And when he wrote the word, I fell in love with it.

Atom.

Atom.

What a beautiful word.

1

Fires of Life

A candle will burn some four, five, six, or seven hours. . . .
Then what becomes of it? Wonderful is it to find that the
change produced . . . is the very life and support of plants
and vegetables that grow upon the surface of the earth.
                                        —*Michael Faraday*

My next breath may very well be in your lungs. Store it
*   wisely, because my life depends on it.*   —*Jarod Kintz*

Take a breath, if you please. You're about to take one anyway, as you've done thirty thousand or so times during the last twenty-four hours. As a newborn you drew your first breaths automatically, perhaps forty times per minute, and even now that you've slowed down closer to half that rate and have the language skills to discuss the process, you still breathe without a thought. You'll die if you stop for too long, and you are hardwired to continue this labor reflexively, even in your sleep.

This time, however, try thinking about it for a moment. Notice how you tighten your diaphragm and relax the muscles in the walls of your chest. This effort alone consumes roughly 3 percent of your metabolic energy at rest, all in order to pull the equivalent volume of a grapefruit into your lungs. Trillions of air molecules are now trapped within your chest like fish in a net. Only a few of them, the oxygens, are what you're after. An average adult uses nearly two pounds of them every day, and this particular breath full will help to keep you alive for the next few minutes. It will also connect you to the rest of life on Earth and to the planet itself in surprising ways that we will soon explore.

People who anticipate having to hold their breath for a long time sometimes try to boost their oxygen stores by breathing purified gas. In 2010 a thirty-eight-year-old free diver named Peter Colat spent nineteen minutes and twenty-one seconds submerged in a water tank in Saint Gall, Switzerland. Shortly before plunging in to become the latest world champion of breath holding, Colat inhaled pure oxygen for several minutes, but the competitive edge that he gained in this manner was only partly due to super-enriched blood. Once your hemoglobin molecules hold their full allotment of oxygen, it becomes difficult to force more of the gas into your bloodstream. Most of the benefit to Colat came from the flushing of stale air from his lungs and the conversion of his inflated chest cavity into a temporary oxygen-storage facility while his mouth and nose were closed.

Your lungs are not necessarily the only route that oxygen takes to your blood, however. You also breathe a little through your eyes. So vital are these oxygen particles that the cells in the transparent surfaces of your eyes absorb them directly from the atmosphere to supplement the meager supply that your blood vessels send to them, as do many of the cells of your skin. And an even more direct way of oxygenating blood was recently developed by researchers at Boston Children's Hospital.

After watching a young girl suffer fatal brain damage before she could be connected to a heart-lung machine, the cardiologist John Kheir sought a way to bypass the lungs and inject oxygen straight into the blood. To do this without producing bubbles that could cause deadly embolisms, Kheir and his colleagues used sound waves to whip pure oxygen and oily lipids into a fine white froth. This trapped the gas in soft, flexible, microcapsules that could oxygenate red blood cells on contact, and when the froth was injected into rabbits they survived on it for fifteen minutes or more without breathing and without obvious distress. "This is a short-term oxygen substitute," Kheir told a reporter for *ScienceDaily* in 2012. "A way to safely inject oxygen gas to support patients during a critical few minutes." If the method can be perfected for human use, it could be deployed in emergency rooms—and presumably, in breath-holding competitions, as well.

For most of us, however, breathing is our primary link to airborne oxygen, and it is so universal and continuous that it can be easy to forget about—until we can't do it anymore. Then it becomes more obviously precious and symbolic of life itself. We take special note of words that are carried on final breaths, and sometimes we even cherish the physical substance of the breaths themselves. Henry Ford kept a small glass test tube of air in his home for many years, and inside the tube was said to be a sample from the last breath of his late friend and fellow inventor Thomas Edison. According to sources at the Henry Ford Museum in Dearborn, Michigan, several such tubes are believed to have been left open to the air of the room near Edison's deathbed. "Though he is mainly remembered for his work in electrical fields," Edison's son Charles reportedly said, "his real love was chemistry. It is not strange, but symbolic, that those test tubes were close to him at the end." After Edison's death, Charles had the tubes sealed and later passed one of them on to Ford as a memento.

But *why* do you breathe? Why do you so desperately need those invisible specks of oxygen? Where do they come from, what do they do inside of you, and where do they go after they leave? You can't even see them, although you live amid dramatic evidence of their presence in the seemingly empty spaces of your world, from rustling leaves and billowing sails to rusting metal and the soft glow of a candle. Until a mere century or so ago, many reputable scientists even doubted their very existence.

Oxygen atoms lurk almost everywhere in your daily life. They are not only things that you breathe but also things that you are largely made of. Scientists sometimes describe human beings as "carbon-based life forms," but a strict tally of numbers says otherwise.

Roughly 60 percent of your weight comes from water, depending on your size, age, and health, and nine-tenths of the mass of a water molecule is dominated by the bulky oxygen atom within it. Therefore oxygen comprises most of your wet mass, and an adult body weighing 150 pounds contains about 95 pounds of it. By comparison that same adult's carbon supply would weigh only 35 pounds and be outnumbered by oxygen two-to-one on an atom-for-atom basis.

In the remaining dry parts of your body, oxygen atoms are also interwoven with other elements in the protein fibers of your tendons, the soap-bubble membranes of your cells, and the coiled tendrils of your genes. Oxygen represents a little more than half of the mass of the blood sugar in a human artery and of the lactose in a drop of human milk, as well as most of the mineral-forming atoms in human bone. If all of your oxygen atoms were to vanish you would still be visible, though not for long. The mist of leftover elements would scatter with the first puff of wind.

But that is not why you breathe. You collect most of your water-bound oxygen atoms by drinking, and your carbon-bound oxygen by eating. Breathing is another matter entirely. The target in this case is not merely oxygen atoms for their own sake, but pairs of them that are joined together in reactive molecules of oxygen gas. And unlike eating or drinking, you have to inhale this gas continuously because, apart from the inflatable bags of your lungs, you can't safely hold much of it inside of you.

Even if you could purify and compress a lot of it into some internal storage space, you wouldn't want to do so. Left unguarded within your body, it can attack and damage your cells, and it is toxic in high doses. You need to consume your oxygen in controlled sips, using it immediately and efficiently, and then take more from the sea of atoms that surrounds you.

The story of how scientists uncovered the atomic connections between us and the air includes a centuries-long sequence of incremental advances and dead ends. It is too long a tale to relate in depth here, but a short summary of some key discoveries during the eighteenth century can at least show that the knowledge was hard-won. Although many investigators made important contributions during that time of intellectual and cultural ferment, the work of three scientists in particular will help to show what can be accomplished with relatively simple but clever experiments.

To set the stage, we enter the eighteenth century armed with two legacies from the 1600s, one misguided and one prescient. The former was a hypothesis by the German alchemist Johann Becher that combustion is simply the release of a mysterious substance called "phlogiston." The latter was the work of Robert Boyle, an Irish-born scientist who proposed that air is not a pure element but a mixture of gases. In his book *Suspicions about the Hidden Realities of Air*, Boyle wrote, "I have often suspected that there may be in the Air some yet more latent Qualities or Powers. . . . For this is not as many imagine a simple and elementary body, but a confused aggregate . . . (and) perhaps there is scarce a more heterogeneous body in the world."

During the early 1770s the Swedish chemist Carl Scheele heated a powdered oxide of mercury and concluded that the "fire air" it produced also exists in the atmosphere, but he did not publish his results quickly. A few years later the British chemist Joseph Priestley did the same, called the emissions "de-phlogisticated air," published the finding before Scheele did, and also showed that the gas could keep a flame burning and a mouse breathing in a sealed glass jar. The French chemist Antoine Lavoisier, having heard of Scheele and Priestley's experiments, conducted similar tests and then claimed that the discovery was at least partially his.

None of these men were in fact the first to create oxygen gas experimentally—alchemists had already produced it during the previous century without fully understanding what it was. But Scheele recognized this vital gas as a distinct component of air; Priestley demonstrated its links to respiration and combustion; and Lavoisier gave it the name we use today. He called it "oxygène" (from the Greek words for "acid maker") because combining it with nitrogen and sulfur in solution yields nitric and sulfuric acids. Along with others Lavoisier also helped to overturn the phlogiston hypothesis. By heating tin in a closed container, he showed that the metal gained rather than lost mass, and that the gain in the oxidized metal matched the mass of the air that rushed in when the container was opened.

The personal lives of these men are also interesting in their own right, and their biographies are full of dramatic details. They struggled bitterly over credit for the discovery of oxygen, and all three met with great troubles later in life. Scheele died at age forty-three in 1786, supposedly from sniffing and tasting too much mercury, arsenic, and lead in his lab. Priestley was driven from England for his liberal religious views and for supporting the American and French Revolutions, and while living in exile in Pennsylvania he endured the deaths of his wife and young son, isolation from the scientific community of Europe, and accusations of sedition. Lavoisier, one of the king's hated tax collectors, was guillotined during the French Revolution, and the mathematician Joseph-Louis Lagrange famously wrote of the execution, "It took them only an instant to cut off his head, but France may not produce another such head in a century."

The discoveries made by scientists such as these led to later investigations on the atomic scale, but without high-tech equipment or scientific training, the invisible nature of atoms still makes it difficult to recognize how oxygen and fire work. It is still easy to overlook, for example, as Priestley and Becher did, the increase in total mass that fire produces in burning fuel, because the buoyancy and dispersal of the waste gases conceals it. Burning a six-pound gallon of gasoline releases about nineteen pounds of heat-trapping carbon dioxide into the atmosphere, and the U.S. Energy Information Administration estimates that transportation vehicles in the United States alone released more than a billion and a half metric tons of CO_2 in 2012. Such perceptual limitations can make it difficult to notice our effects on our surroundings, and they can sometimes mislead us about the atomic nature of the air we breathe.

After watching mice respire easily in a jar of fire air nearly two and a half centuries ago, Priestley sampled some of the gas for himself. "The feeling of it to my lungs was not sensibly different from that of common air; but I fancied that my breast felt particularly light and easy for some time afterwards. Who can tell but that, in time, this pure air may become a fashionable article in luxury?" Who, indeed, could have foreseen that fashionable people of the twenty-first century might pay a dollar a minute to pump oxygen gas into their nostrils through plastic hoses?

First developed in large, pollution-plagued cities such as Tokyo and Beijing, "oxygen bars" spread during the 1990s to other cities around the world and have now spawned an additional market in portable oxygen dispensers for home use. Proponents say that the gas removes toxins from the body, strengthens the immune system, cures hangovers, and performs other medical miracles, but little or no empirical evidence supports most of these claims. As George Boyer of Mercy Medical Center in Baltimore explained in an interview for WebMD, "If your lungs are healthy, and you have no breathing difficulties, your body has all the oxygen it needs," adding that "for the vast majority of people there is little harm [in using an oxygen dispenser], but also absolutely no science of benefit."

Whether you buy your oxygen in a bar or inhale it for free, the question remains: Why do you do it at all? Many early investigators thought that breathing was done merely to cool the "animal heat" of one's body, and that lungs were little more than air conditioners. The reality lies deep within the cells of your body, where the differences between combustion and respiration are also more clearly revealed.

The basic formula of "food plus oxygen yields CO_2 and water" that is often presented in textbooks seems to imply a direct connection between the gases that enter and leave your lungs, as though the metaphorical fires of life operate on the atomic scale precisely as real fires do. Many scientists foster this impression, sometimes to simplify concepts for nonspecialists, and sometimes because similar explanations led them astray earlier in their careers. A recent article on molecular physiology in *Science*, for instance, described how blood sugar is "burned" with oxygen, and college professors often suggest in their lectures that the carbon dioxide we exhale is an exhaust gas that forms when breath oxygen combines with sugary fuel in the furnaces of our cells.

The image is appealing, but it is also wrong. A closer look at the subject shows why fire doesn't perfectly illustrate your use of oxygen.

Fire does resemble a living thing in many ways. Both emit carbon dioxide and—strangely enough—water vapor. Although a rush of liquid water can extinguish a flame or drown a person, the gaseous form of water in a puff of smoke or a human breath has no such effects on its source. Flames and life are also alike in that both can be snuffed out if deprived of oxygen. The light and heat that emerge from a candle arise from the breaking of chemical bonds in wax molecules, and the warmth of your skin is related to the breaking of bonds in food molecules. But although the basic equations of combustion and respiration are similar, the processes at work in a fire differ from those that sustain you.

In a candle flame, oxygen gas from the surrounding air attacks the melting wax directly, tearing electrons away from fuel molecules in a swirl of glowing carbon-rich particles and partially ionized gases. When the gaseous body of a flame is hot, dense, and ionized enough, as for example on the six-thousand-degree business end of an oxy-acetylene torch, it is called a "plasma," a term that also applies to the incandescent sphere of the sun. Plasma is the fourth state of matter, a dynamic complement to the more commonly recognized solid, liquid, and gas states, and it may well be the dominant readily visible form of matter in the universe because stars are made of it. Here on Earth, less ferocious kinds of fire unwind carbon-based fuels into simpler particles much like those from which they originated. For instance, the petroleum from which candle paraffin is made was built from carbon dioxide and water by light-harvesting algae. Burning a candle returns its carbon and hydrogen atoms to the atmosphere in the grip of oxygen as carbon dioxide and water molecules. It also unleashes the solar energy that originally bound those raw materials together into living tissues, and it does this so rapidly that temperatures inside the hottest parts of a candle flame can reach 2500°F.

Inside the controlled confines of your cells, however, oxygen gas is usually not so much a fierce lion as a well-trained house cat that waits to be fed. When hot wax breaks down in a candle flame, for example, oxygen gas swoops into the blaze and emerges with carbon and hydrogen atoms in its clutches. When carbon-rich food oxidizes in your body, those same two waste products (CO_2 and H_2O) are made in two separate processes. The oxygen gas that you inhale will not carry carbon atoms off in later breaths as it might if you were a candle. In the relatively tame habitats of your cells, oxygen specializes in capturing hydrogens instead. And fortunately for you, the production of energy by cellular respiration is slow and dispersed enough to warm you without immolating you.

To envision how this process works, it helps to use Albert Einstein's technique of conducting "thought experiments" to let your imagination follow some oxygen molecules down your throat and into your lungs on your next breath. About three-quarters of the air that you have just inhaled consists of nitrogen molecules, none of which will serve you apart from helping to keep your lungs inflated. Your target is the 21 percent of air that consists of oxygen molecules, but you can only get at them by harvesting the whole mess, like scooping up piles of mixed jelly beans and then picking out the colors you prefer.

As your chest expands and air presses into it, the gas squeezes through bronchiole ducts as thin as a human hair into hundreds of millions of bubble-like alveoli that compose the pink spongy interiors of your lungs. Their combined absorptive surface area approaches 750 square feet, roughly equivalent to one-third of a singles tennis court. From there most of the air molecules work their way into narrow spaces between the alveoli, where blood capillaries gather them up. On that microscopic scale your blood resembles water crammed with translucent blobs of crimson Jell-O that can squeeze past an alveolus within less than a second as your lungs pulsate. Those red blood cells are fast-moving vehicles that can carry oxygen through hundreds of miles of vessels to destinations all over your body.

Meanwhile, carbon dioxide molecules that recently formed inside your cells stream out of your blood and into the alveoli. In all that confusion most of your newly inhaled oxygen is simply blown right back out of your lungs. Although wasteful in one sense, such inefficiency can be a good thing. Residual oxygen allows mouth-to-mouth rescue breathing to revive an unconscious person rather than asphyxiating them with carbon dioxide.

A pint of your blood can carry roughly one-fifth of a pint of oxygen gas, almost enough to sustain you at rest for one minute. But the supply runs lower along the journey from your lungs to your cells and back. By the time venous blood makes it back to your alveoli, it contains so little oxygen compared with the air in your lungs that the imbalance automatically drives the diffusion of more oxygen into your blood.

The primary goal of an oxygen molecule in your body, if molecules can be said to have goals, is to be dismembered inside of you. But if you could accompany it on its fatal trip to one of your cells, you would have to do so as a purely metaphysical being because, naturally, you couldn't be made of unshrinkable atoms and still be yourself, any more than a brick building could shrink to the size of a standard brick and still be recognizable. Scaling the atomic realm up to match you would not work either, because everything around you in that fantasy world would then have to be moving unrealistically fast. An atom is about ten billion times smaller than you are, and an oxygen atom's trip from your heart to your hand would be millions of miles long on an equivalent human-size scale. The blood in the brachial artery of your arm covers this distance within a single second, so a person-size atom would have to travel faster than the speed of light to do the same, which Einstein's research on relativity showed to be impossible.

Even without the improbable logistics of shrinking and growing involved, the atomic realm is much stranger than the ones most of us are used to. Atoms are so mobile and the outer boundaries of their electron clouds so unstable that the atomic surfaces of objects are more indistinct than firm. In that world of the incredibly small there is no air to breathe, no sound to hear, and visible light cannot illuminate objects as it does in our much larger size range.

Nonetheless, here is an example of what you might imagine seeing and feeling in a thought experiment if a single skin cell were magically inflated ten million times into a living hill three hundred feet high so you could more easily see what happens to oxygen inside of it. At that magnification the sizes of the atoms making up the cell approach those of sand grains, and the body that you normally occupy would be large enough to lie down with your head in New York, your waistline atop the Pacific Ocean, and your feet in Australia.

You will now need to enter that hill by forcing your way in through the pliant, oily membrane. It is wet in there, so suspend disbelief even further and assume that you can still breathe despite the syrupy substance that fills the cell. The scene looks positively industrial. Structural protein cables as thick as your arm stretch in all directions, giving the cell its shape.

Just over there is your destination—a cylindrical bubble-like thing that is roughly the size of a tractor trailer truck. This is a mitochondrion, a living power plant that uses food as a fuel. Each of your cells may contain dozens to hundreds of these, and they can vary in shape from peas to noodles. It is within such mitochondria that your breath oxygen meets its doom.

Enzymes in the cell and the core of the mitochondrion smash food molecules into a rich stew of electrons, hydrogen ions, and CO_2. The electrons are then fed to a series of proteins that lie embedded in a soft membrane surrounding the core, some of which will twitch, bend, or roll as the electrons pass through them. And as the molecular machinery churns, it also stores chemical energy that can power muscles and metabolism. In other situations it can help to generate body heat instead.

At the end of the line, each exhausted electron makes one last jump to clear the way for the ones coming up behind it. And here is the precise point from which your need for oxygenated air arises.

Oxygen uses those leaping electrons to tether the hydrogen ions from shattered food molecules to itself. In this transmutation of food and air, disparate components of your meals and breaths recombine to create H_2O molecules, your own homemade metabolic water. A tenth of the fluid in your body, from the blood in your veins to the moist gleam in your eyes, is built from scratch this way with the aid of oxygen that you inhaled during the last few days. Air and water are therefore more closely related to one another than the alchemists ever imagined, as each can represent a reconfiguration of the atoms of the other.

This, then, is why you breathe. You stoke the machinery of your cells with air, carry the watery leftovers around for a time, and then release them back into your surroundings through sighs, sweat, tears, and more substantial wastes. In doing so you split a flame's oxidation of carbon and hydrogen into two separate processes, thereby creating situations in which only metaphorical fires of life will flicker. The atmosphere literally becomes a part of you every time you draw a breath, and part of you returns to it on every outgoing breath as well.

Depending on the time of day and the season of the year, the air you walk through and pull into your lungs changes more than you might expect. This is just one of many discoveries by Ralph Keeling, a scientist at the Scripps Institution of Oceanography who tests the atmosphere the way a police officer might test your breath with a Breathalyzer.

For more than two decades Keeling has been measuring the oxygen content of air samples that are collected daily in Hawaii, Antarctica, and elsewhere, sealed into small containers, and shipped to his lab in La Jolla, California. Like traces of alcohol in someone's breath, slight changes in the composition of the atmosphere can tell a lot about what the world's combined masses of people, vegetation, and plankton are doing.

It is often said that forests are the "lungs of the planet" because they produce oxygen that we breathe, but the metaphor falls short in some respects. Lungs don't produce oxygen but instead consume it, and Keeling's work has shown that only about half of your oxygen comes from terrestrial plants. The rest is made by algae and cyanobacteria in lakes and oceans, with a small additional measure produced by the splitting of water vapor in the upper atmosphere by radiation from the sun and distant stars.

However, when combined with the carbon dioxide analyses that his late father, Charles David Keeling, launched at Mauna Loa Observatory in Hawaii in 1958, the long-term oxygen records do show an almost eerie resemblance to the readouts of a medical breath-monitoring device. Annual pulses of oxygen are mirrored by cyclic drops in CO_2, and together these data open a unique window on the atomic connections between plants and the earth.

When the elder Keeling first began to study the air, he expected it to vary a great deal from place to place. To his surprise, however, much of the variability vanished when samples were collected with consistent methods at remote locations where the air is free of local influences from respiring forests and cities. The atmosphere mixes more thoroughly and rapidly than scientists had hitherto realized, and average CO_2 concentrations in Hawaii are remarkably similar to those at the Scripps pier in La Jolla.

Equally noteworthy, however, were various kinds of rhythmic oscillations that appeared in the gas records. Every day the carbon dioxide concentrations dropped slightly, only to recover at night, and larger seasonal pulses occurred with dips in summer and peaks in winter. When Ralph Keeling began to measure oxygen to complement his father's work, his results showed similar patterns but in reverse. With these data you can watch the atmosphere respond to the breathing of countless plants and microbes as the earth spins on its axis and circles the sun.

The pacemaker of these pulses is sunlight. When dawn awakens California, the lawns and palm trees of La Jolla begin to pump oxygen into the air and pull carbon dioxide out of it, as does the Pacific plankton drifting offshore. When that portion of the world spins onward into the shadow of night again, the oxygen production shuts down, but the cellular CO_2 factories, which need no solar power supply, keep running and quickly drive local carbon dioxide levels back up again while oxygen levels drop.

A similar pattern emerges in alternating hemispheres through the seasons, as well. When plants sprout and leaf out in spring, O_2 rises rapidly and CO_2 declines. Later in the year when photosynthesis slows and dead leaves begin to decay and release carbon dioxide, the opposite trends prevail. This pattern produces a sawtooth effect in the Keeling charts, and if you could hear the oscillations rather than see them they would sound like the breathing of a long-distance runner. The oxygen curve would go, "Puff hard in spring, take a long deep breath in winter, puff again," and so on.

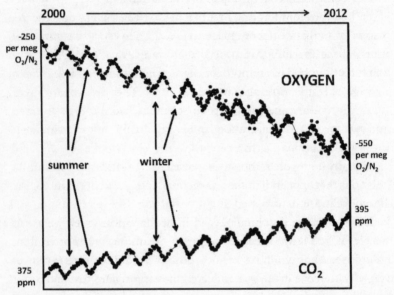

Seasonal cycles and trends in the oxygen and carbon dioxide contents of the air at Mauna Loa between 2000 and 2012. The overall decline in oxygen is largely the result of fossil fuel combustion along with fires and decay associated with land clearance. *Data courtesy of the Scripps O$_2$ Program*

Here again, however, the lungs-of-the-planet metaphor has limits. Not only do these lungs emit oxygen; they do it asymmetrically. When one hemisphere exhales the other one inhales, which would be interesting to watch if real lungs were to do so within a runner's chest.

The Keeling records clearly show that we affect the atmosphere, too, but in more disturbing ways. In early 2013 average concentrations of heat-trapping carbon dioxide reached 400 parts per million (ppm, or a ten-thousandth of a percent), having risen from an average closer to 312 ppm during the 1950s. Most of that change represents the burning of fossil fuels along with the decay and fires associated with deforestation. Unlike the photosynthesizers, these artificial "lungs" of the modern world consume O_2 and release CO_2 like our own, and they do it continuously on a massive scale.

While the long-term carbon dioxide record tilts upward along with global average temperatures, the oxygen trend points downward. According to the Scripps O_2 Program Web site, oxygen concentrations at La Jolla dropped by 0.03 percent between 1992 and 2009. This, as Ralph Keeling said in an interview with the *San Diego Union-Tribune*, is the global "signature of combustion."

Should we now worry about running out of oxygen in addition to global warming? Not according to Keeling. In another *Union-Tribune* interview he explained that there is plenty of oxygen in the air, and the tiny percentage of loss of oxygen in itself isn't an issue. Rather, "the trend in oxygen helps us to understand . . . what's controlling the rise in CO_2."

In other words, declining oxygen shows how closely tied we are to this planet, and how much we now affect the atomic world around us. The Scripps O_2 Program reports that roughly a trillion metric tons of oxygen were consumed by fossil fuel combustion since the Industrial Revolution, but it still amounts to a mere tenth of a percent decline in the huge oxygen reservoir of the atmosphere. This is far too small to have any direct impact on human health, and seasonal oxygen levels can often drop by 10 percent or more over major cities without ill effects. So enormous is Earth's oxygen reservoir that the annual output of all the world's plants and plankton barely dents it. A million billion metric tons of oxygen molecules ride the air, representing about two thousand years of seasonal contributions. During cooler, darker seasons and at night when photosynthesis shuts down, life survives mainly on oxygen that was emitted long ago.

To bring this concept down to a more personal level, you could probably walk over and touch some of the kinds of oxygen sources that are helping to sustain you right now. A study published by the research forester David Nowak and his colleagues in *Arboriculture & Urban Forestry* suggests that an average acre of trees creates enough oxygen from year to year to keep eight people alive, although the exact relationship varies according to the species, age, and environmental settings of the trees. American urban trees as a whole produce a net total of 67 million tons of oxygen per year, enough to sustain two-thirds of the population of the United States. Of the United States cities listed in that report, the champion oxygen generator was lush and leafy Atlanta, Georgia, which could keep most of its residents perpetually breathing with the 95 thousand tons of oxygen that its own trees release each year. New York City and Washington, DC, came in second and third with 61 thousand and 34 thousand tons, respectively. Freehold, New Jersey, came in last with only 11 hundred tons.

There are many fine reasons to plant trees in cities, from temperature control to aesthetics, but oxygen production is not one of them. You can always find plenty of perfectly good oxygen to breathe in the most barren desert or, for that matter, in Freehold, New Jersey. And no matter where you live, you always have more than enough oxygen available during winter even though the local trees are leafless or dormant. Most of your oxygen molecules come from more distant places and times, and only a minute fraction of the oxygen in your next breath was produced during the past year.

Although it can be tempting to think that plants make oxygen gas for your benefit, it isn't so. What you breathe is merely the accidental crumb droppings of a feast that light harvesters prepare for themselves. Plants use oxygen like you do, and they consume almost as much of it as they produce. Any net production of oxygen by a forest or ocean is simply leakage, and much of what does escape is soon consumed in rust, decay, and fire. If tons of dead plants, animals, and microbes were not continuously buried in soils and ocean sediments before they could rot or burn, the atmospheric oxygen account would eventually balance out to zero and everyone would suffocate.

Your other oxygen sources can be found with the help of a fine-mesh net that you might pull through a patch of seawater. If the haul is mostly yellow-brownish, then you have probably harvested single-celled algae known as diatoms. Under a microscope they look like dabs of golden jelly encased in glistening glass shells that can look like snowflakes, needles, or designer hubcaps. Other kinds of photosynthetic plankton bear shells of creamy white chalk or resemble strands of green malachite beads. Plant-like relatives, the seaweeds, can resemble green cellophane, long sheets of soft brown rubber, or crinkly lettuce. Ralph Keeling has calculated that marine photosynthesizers such as these produce nearly thirty billion tons of oxygen per year in the Northern Hemisphere, while the larger southern oceans yield roughly half again as much.

Oxygen shortages are most likely to become a problem if you move too far upward through the sea of air. Gravity pulls gas molecules downward and crowds them more tightly together at ground level, and at sea level you operate under an average pressure of roughly fifteen pounds per square inch. The weight doesn't normally bother you, though, because the air is evenly distributed over your body and the pressure inside you matches that on the outside. It is more like floating in a swimming pool than like having three-quarters of a ton of water balloons piled atop your head and shoulders. Most of us don't even notice this ever-present pressure until our ears pop in an airplane or on a road trip through tall mountains.

Visualizing air pressure. Left: A plastic bottle sealed shut at about 15,500 feet altitude in the Peruvian Andes, where there are half as many air molecules per breath as at sea level. Right: The same sealed bottle being squashed by the pressure of more densely packed air molecules near sea level in Lima. *Photos by Curt Stager*

Atop a mile-high peak, you still get 83 percent of the oxygen that you would at sea level, according to an air-pressure calculator posted online by altitude.org, but if you climb beyond eight thousand feet or so where relative oxygen abundances in the thinning air drop to 75 percent or less, altitude sickness could begin to present problems. People sometimes develop headaches in high-flying commercial airliners, and in some cases this may reflect mild hypoxia due to low cabin pressure, which can resemble conditions on a mountainside at a mile-and-a-half elevation.

People whose ancestors lived at high altitude for many centuries tend to carry genetic mutations that can help them to cope with low oxygen supplies. Many Tibetans breathe faster and take in more air per breath than the rest of us do, a genetic inheritance that helped their ancestors to make up for the shortage. Among many Andeans, large amounts of modified hemoglobin in the blood also extract more oxygen per breath.

But even for such specially equipped people, long-term existence on the world's tallest crags is impossible. We simply can't afford to stretch our atomic connections to photosynthetic life too far.

Careful research has shown that we breathe in order to harvest oxygen gas from the air, that we convert it to water, and that this need for oxygen somehow links us to plants. But what, exactly, is that vital relationship between plants and people? In order to trace it in more detail, we must focus less on oxygen molecules than on oxygen atoms.

Consider the atomic connections between you and a potted plant. If you and your plant were to spend a few hours together in a well-lit room, then you might be able to play a game of catch with a single oxygen atom in different molecular forms. You couldn't use your breath CO_2 for this whimsical thought experiment, though, because the plant would turn it into vegetable matter, and your main method of getting an oxygen atom back after releasing it in a CO_2 molecule would be to eat the plant.

If instead you were to inhale an oxygen molecule that emerged from the plant, your cells could turn it into metabolic water. Now separated and "disguised" as H_2O, one of the oxygen atoms from that original gas molecule might escape through your breath and drift back through the air to the plant in the form of water vapor. If that molecule were to enter a leaf it might be smashed in photosynthetic reactions that could spit the same oxygen atom back out at you again in yet another oxygen gas molecule.

Supported by many more scientific discoveries than Scheele, Priestley, and Lavoisier had access to, you can now use your well-informed imagination to follow the botanical side of this exchange even deeper into the cellular level. Plants turn water into oxygen with the help of green chloroplasts that somewhat resemble mitochondria in size and shape but are packed with layered membranes that resemble spinach lasagna. When sunlight strikes a leaf, some of it hits emerald-colored molecules that are anchored in the chloroplast membranes. Those sun-struck chlorophylls, in turn, fire electrons into molecular machines that help to drive the construction of sugars which can later become sap, stems, flowers, and seeds.

But there's a hitch. The chlorophylls must be reloaded after firing. In a leaf, water molecules are the handiest sources of those electrons, and chloroplasts excel at ripping water apart to get at them. Solo oxygen atoms that are left over from that demolition join to form gas molecules that may be consumed immediately in nearby mitochondria or released into the air, perhaps eventually to visit your lungs.

The broad patch of sunlight that slides across the face of the earth over the course of a day triggers great bursts of photosynthesis wherever it goes. If water molecules were larger, you might hear them exploding like firecrackers as solar-powered life smashes them and sprays molecular shrapnel into the air and oceans. The oxygen gas that emerges from this mayhem faces many possible fates, all of which eventually end in destruction whether it be within a fleck of rust, a flash of lightning, or a cell inside your fingertip.

But if oxygen molecules were sentient beings, they would surely take such things in stride. On a planet such as this which is so full of life, the demise of any given molecule is as inconsequential in the longer history of its component atoms as it is inevitable. By some estimates, today's rates of global oxygen production could theoretically smash every water molecule in the oceans within several million years, and in their book *Plant Physiology*, Hans Mohr and Peter Schopfer calculated that land plants alone have split the equivalent of Earth's entire supply of surface water about sixty times since the evolution of large forests four hundred million years ago.

From a geological perspective, then, much of the water and most of the oxygen gas that you use is relatively young, with its age more likely counted in mere centuries or millennia than in millions of years. Therefore, although it can be tempting to think that what you drink and breathe today could be the very same stuff that dinosaurs once used, most of it probably isn't.

On the other hand, the atoms from which those air and water molecules arise are much older, and they not only connect you to the atmosphere, oceans, and organisms of the earth—your atomic connections also extend into the depths of space and time as well.

The water that moistens your mouth and flows through your veins represents multiple generations of atoms: hydrogen from the birth of the universe and oxygen from the hearts of stars. Shortly after the Big Bang 13.8 billion years ago, your hydrogen atoms condensed amid clouds of subatomic particles. There was no oxygen at all then because oxygen atoms form within stars, and the first stars had not been born yet. The oxygen atoms that now reside within you are therefore younger than your hydrogen atoms, having formed millions or even billions of years later.

The first stars ignited when primordial hydrogen formed clumps that became large, dense, and hot enough to trigger nuclear fusion reactions. Similar reactions also power the sun, whose core temperatures are measured in tens of millions of degrees. In that painfully brilliant plasma, electrons dash away from their host nuclei, which then crash directly against one another without the protection of electron shells. So violent are the collisions that they overcome the mutual repulsions of positively charged protons and allow more powerful nuclear forces to click them together. These new clusters form the nuclei of helium atoms, which are fittingly named after Helios, the ancient Greek god of the sun. This gigantic solar fusion reactor releases so much energy that it drives the entire photosynthetic activity of the earth while also baking you on a beach towel from 93 million miles away.

Most of your oxygen atoms, however, were probably born within larger stars that fused their hydrogen nuclei into much larger elements. The nuclear ancestors of the elemental oxygen that gives your body water most of its mass were once identical to the older, smaller hydrogen atoms that now perch upon it. When the megastars matured, senesced, and died, your newborn oxygen atoms blew off into space like pollen from bright, flaming flowers. To look into the night sky is to survey distant gardens in which the elements of life are ripening, and your body is a composite harvest from those cosmic fields. Throughout history, people have spoken of the earth as our mother and the sun as our father, perhaps reflecting prevailing views of traditional sex roles. In an atomic sense, however, it would be more accurate to think of the earth and the sun as our siblings, because they both formed from the same star debris as the elements of life within us. Earth is indeed a kind of surrogate mother to us in that our bodies are derived from it, but we exist today only because our true celestial star mothers died long ago. May your next breath of fresh air be dedicated to their memory.

From space, Earth resembles a floating blue bead, and if you keep that image in mind it will help to drive home a lesson that was arguably one of the most important contributions of the NASA space program. As abundant as atoms are on this planet, their numbers are finite. Watch a satellite video of the clouds that sweep across the face of the world, and you will see in an instant that the winds that carry them over one curved horizon may reappear on the opposite horizon: The truth of such seeming platitudes as "What goes around comes around" becomes obvious.

When viewed from a great distance the sky resembles a shockingly thin film, and most of its molecules are packed into a mere ten-mile slice of a total planetary diameter of nearly eight thousand miles. At sea level you might find more than ten trillion trillion atoms in a cubic yard of air, but just outside that vaporous skin is the relative vacuum of the solar system, in which fewer than a dozen atoms might enter each of your lungs should you try—and fail—to breathe out there. The next time you see a photo of the earth taken from space, try to convince yourself that a pollutant-spewing smokestack anywhere in the world doesn't unleash potentially harmful substances into the same precious air supply that keeps you and your loved ones alive.

Keeling and his colleague Stephen Shertz showed that oxygen gas emitted by plants and plankton mixes throughout each respective hemisphere within two months and spreads worldwide in a little more than a year. The sensitivity of the oxygen and carbon dioxide balance of the atmosphere to the activities of living things shows that recycling is not just a passing fad but a tradition that has always been practiced on the atomic level by all life on Earth. To live, rather than to merely exist like inanimate rock, is to borrow and repurpose the elements of the world around you, and then release them again.

The first time I encountered this line of thought during the early 1960s, I was reading a comic book when I should have been doing my homework. I don't recall the title or even the main point of the story, but one cartoon panel still stands out in my memory. In it a gray-haired inventor hunched over a workbench, and the caption read, more or less: "You are now breathing some of the air molecules that Leonardo da Vinci himself breathed during his lifetime."

Being a young kid, I didn't fully grasp the finer points of that story. I didn't even understand why Leonardo was the main character. Was it because he breathed more air than the rest of us? But many years later, having learned that Leonardo is merely one among several historical figures who are commonly used to illustrate breath recycling (including Julius Caesar, Jesus, Shakespeare, and Hitler), I still find it intriguing. So do many other people, judging from the large number of hits in a Web search of the subject.

You have to pick carefully through the facts and figures necessary to make such a calculation, but most sources end up reporting the presence of between one and fifteen breath atoms from the person in question within each of your own breaths. To begin with, a reasonable value for the total mass of the atmosphere is needed, such as the five million billion metric ton estimate that was published in 2005 by Kevin Trenberth and Lesley Smith of the National Center for Atmospheric Research in Boulder, Colorado. That represents something like 10^{44} air atoms (that is, unimaginably many), of which a fifth are oxygens.

If you also take into account gas exchange with the oceans, the volume of an average lung, the breathing rate of an average human, and so on, you're likely to come close to the range of one to fifteen Leonardo atoms per breath and reach the same basic conclusion: We all share the same air. But to make your calculation as accurate as possible, you must also select the right kind of gas to follow.

The one-to-fifteen estimate doesn't apply to oxygen gas, for example, because it is too unstable to linger in the air for very long. Cells, forest fires, lightning bolts, and space radiation are likely to smash it sooner or later, turning it into water and other compounds. In a crowded elevator you do share some oxygen molecules with your companions, as the waning fraction of unused air in your lungs passes from person to person. But on a planetary scale, the chance of your sharing an oxygen molecule with Leonardo or a more distant historical figure is closer to nil. Most of the oxygen that you breathe is too young for that. And although nitrogen gas is much more abundant and stable, it nevertheless cycles through food chains and can therefore leave and return to the atmosphere as well.

A better gas to illustrate the recycling of air would be one that isn't created or consumed by living things. The most widely cited choice is that of the American astronomer Harlow Shapley, who used argon as the subject of his calculations.

Argon is ubiquitous in the atmosphere, although it represents less than 1 percent of the total. That rarity is a good thing, because breathing high concentrations of argon is a bad idea. It is much heavier than most gases and therefore harder to force out of your lungs once it enters. Suffocation by argon in industrial accidents is more common than death by more overtly poisonous gases such as chlorine. But 1 percent is nothing to worry about, and it makes a convenient air tracer for Shapley's thought experiment, which yielded a high end estimate of fifteen argons per breath that are likely to have been inhaled previously by someone like Leonardo.

His 1967 essay, "Breathing the Future and the Past," followed the argon atoms in a single breath so vividly that I'll quote it here:

We shall call it Breath X. It quickly spreads. Its argon, exhaled this morning, by nightfall is all over the neighborhood. In a week it is distributed all over the country; in a month, it is in all places where winds blow and gases diffuse. By the end of the year . . . [it] will be smoothly distributed throughout all the free air of the earth. You will then be breathing some of those same atoms again. . . .

This rebreathing of the argon atoms of past breaths, your own and others', has some picturesque implications. The argon atoms associate us, by an airy bond, with the past and the future. . . . You contribute so many argon atoms to the atmospheric bank on which we all draw, that the first little gasp of every baby born on Earth a year ago contained argon atoms that you have since breathed. And it is a grim fact that you have also contributed a bit to the last gasp of the perishing.

Every saint and every sinner of earlier days, and every common man and common beast, have put argon atoms into the general atmospheric treasury. . . . Argon atoms are here from the conversations at the Last Supper, from the arguments of diplomats at Yalta, and from the recitations of the classic poets. . . . Our next breaths, yours and mine, will sample the snorts, sighs, bellows, shrieks, cheers, and spoken prayers of the prehistoric and historic past.

Shapley was no poet or mystic, and his nerd credentials were impeccable. He studied astronomy at Princeton, helped to establish the National Science Foundation, served as director of the Harvard Observatory, and calculated the size of the Milky Way. But I'm not surprised that he waxed so eloquent in his essay, or the least bit shocked by the whiffs of philosophy and spirituality in it. Strong undercurrents of emotion often emerge amid the facts and figures that science uncovers, and when discoveries are truly profound they can release profound insights and feelings, as well.

As brilliant as he was, Henry Ford apparently failed to realize that he needed no test tube to capture the atomic essence of Edison's last breath. You can collect a sample of it anytime—along with samples from the last breaths of Caesar, Jesus, Shakespeare, Hitler, and Leonardo—and even with a few bits of air that carried your own first cries as a newborn.

It's easy to do, here on this sky-blue sphere of atoms. Just take a breath, if you please.

2

The Dance of the Atoms

*The atoms are dancing. . . . All the atoms in the air and in
the desert, let it be known, are like madmen.*

—*Rumi*

If there is magic on this planet, it is contained in water.

—*Loren Eiseley*

You cross a bridge and marvel at the strong beams that support it but
overlook the little bolts that also hold it together so you (and they)
don't fall. A bit of grit in your salad grabs your attention more vividly
than the softer parts of the mouthful in which it hid. And after you've
had nothing to drink for a long time, a plain glass of water sure tastes
good. Sometimes the simplest things in life matter the most. The
same holds true for the basic atomic components of your body, without which you couldn't exist. And of those the most basic of all is
hydrogen.

Hydrogen is the essence of simplicity, a single proton accompanied by a single electron, and as with all atoms, most of it is empty
space. But this humblest of elements is crucial to you and the world as
you know it. Without hydrogen, water would no longer exist ("hydrogen" refers to the creation of water from hydrogen and oxygen), so all
the world's oceans, the clouds, the polar ice caps, and 60 percent of
your body would disintegrate. Your muscles would unravel into use-
less tangles of carbon fibers, your bones would crumble, and your
cells would melt for lack of membranes. And all this would occur in
frigid darkness because the sun and stars would disappear. But your

need for hydrogen runs even deeper. It is also the ancestor of the other elements of life.

A friend of mine likes to say that "hydrogen, given enough time, becomes people." That claim, in the grand scheme of things, is correct. Hydrogen first appeared shortly after the Big Bang and ignited the first stars, so hydrogen is the oldest form of atom in your body as well as in the rest of the universe. Your oxygen atoms, on the other hand, can be traced back to the fusion of hydrogen nuclei in later generations of stars. The hydrogen atoms in your water molecules are therefore equivalent to the aunts and uncles of the younger oxygen atoms they ride upon.

But my friend's saying is also true in a more immediate sense. Trillions upon trillions of hydrogen atoms are helping to hold you together right now, and many others are abandoning airborne vapor molecules to snuggle into the atomic frameworks of your hair. Similar connections also link the feathers of desert birds to cacti, the tail bristles of elephants to raindrops, and the hair on the heads of ancient mummies to local drinking water. We'll soon see how that can be, but first you should know something else about the atomic trios that we call water molecules. You might say that they dance.

During the summer of 1827, a Scottish botanist named Robert Brown plucked some pollen from the center of a purple *Clarkia pulchella* blossom and mixed it with a drop of water. Peering through his brass microscope, Brown noticed something odd. Tiny bits of matter in the pollen smear seemed to tremble as they floated in the wet medium under his lens.

Quite naturally Brown assumed that the vibration arose directly from the debris itself. He wrote in his notebook that it "arose neither from currents in the fluid, nor from its gradual evaporation, but belonged to the particle itself." Like many scientists of his time, as well as the philosopher-scientists of ancient Greece who inspired them, he believed that organic matter contains a mystical form of energy that imbues it with life. Perhaps this was the vital essence in action before his eyes!

Being a meticulous scientist, however, Brown tested his hypothesis by examining other kinds of pollen and spores, within and around which he noted twitching flecks of detritus in "vivid motion." To his surprise, inanimate substances also trembled thus in suspension just as the biological samples did. Asbestos fibers, powdered metals, and even a finely ground sample of stone from the Sphinx did the same thing. What these samples all had in common was not a magical life force but the liquid that surrounded them.

We now know that the movement did indeed arise from tiny particles, but those particles were many thousands of times smaller than the ones that Brown could have seen through his simple microscope. He called his trembling flecks "molecules" according to the fashion of the day, and since the translation of "mole-cule" is "small bit of stuff," his terminology was technically accurate. What he really saw, however, was a violent assault on the chunks of matter by the invisibly tiny molecules of the water they floated in. He was also, in a way, seeing himself. The same molecular turmoil churns within all of us and, like the mythical vital force, helps to keep us alive.

Robert Brown's equipment could not show him the true depths of the chaos at work on that submicroscopic scale. The particles he examined were just barely visible, perhaps on the order of a micron or so in diameter, or one-thousandth of the thickness of a fingernail. The water molecules that buffeted them were ten thousand times smaller, even smaller than the wavelengths of light that make things visible to the eye. Today far more powerful microscopes use electrons rather than light to produce vivid images of molecules and atoms, but they don't actually photograph such objects, which lie beyond the normal limits of vision. They merely poke at them to determine their positions and shapes, rather like a phonograph needle that reads the grooves of a vinyl record. And the small sizes of molecules and atoms are not the only barrier to observing them. Equally problematic is their unwillingness to sit still.

When a team of IBM researchers produced the first stop-action film of individual molecules in 2013, they had to chill the sample to −450°F in order to slow their subjects down enough to manipulate them. Now listed in the *Guinness Book of World Records* as the world's smallest stop-motion film, "A Boy and His Atom" features beadlike carbon monoxide molecules that were arranged into the shape of a stick-figure human with a single dot for a toy. Those sharply focused geometric patterns would be impossible to produce at room temperature, because the jiggling molecules would have fled the field of view at speeds close to a thousand miles per hour.

The tremendous velocities of free-floating molecules and atoms make them smash rapidly and repeatedly into their equally energetic neighbors. At room temperature and at sea level, an oxygen molecule collides with its atmospheric companions more than a billion times per second, and if you could hold your hand perfectly still for a single second in what appears to be motionless air, it would nonetheless be pummeled by more than a trillion trillion such impacts. The seemingly empty spaces around you are so full of careening molecules that, if you could clearly see them in their multitudes, it might make you queasy to inhale or walk through them. But you yourself are never completely still either, even when you sleep, because your own atoms and molecules are also madly crashing into one another or straining against the tethers of their chemical bonds.

The molecules that surrounded Robert Brown's samples revealed their presence indirectly by battering the chunks from all directions. Slight differences in the numbers of molecules striking different sides of each particle at any given moment produced a random movement much like a beach ball might exhibit as it bounces atop a closely packed crowd of dancers.

Brown was not the first person to notice the motion that now bears his name. The term "Brownian motion" refers most specifically to the movement of tiny particles in water, but it also reflects a more general phenomenon, the thermal vibration of all atoms and molecules. You have glimpsed something like it yourself if you have ever watched dust motes sparkling in a shaft of sunlight. The glitter comes from flashes of light that bounce away from suspended flakes and flecks as they tumble amid unseen multitudes of dancing air molecules.

Two thousand years ago the Roman poet Lucretius proposed a mechanism for such aerial dust displays. In the classical scholar John Selby Watson's translation of "The Dance of Atoms," Lucretius explained it thus: "Such disorders signify that there secretly exist tendencies to motion also in the principles of matter, though latent and un-apparent to our senses . . . thus motion ascends from the [atoms] and spreads forth by degrees so as to be apparent to our senses."

Lucretius and other early philosopher-scientists speculated on the atomic nature of existence, though they had no way of confirming their hunches. Sometimes their imaginations led them astray, as when Lucretius wrote that atoms are "solid through and through," and today we know that the sparkling of dust motes is more the result of air currents than of molecular vibrations. But as Stephen Greenblatt explained in his book about Lucretius, *The Swerve*, these early atomists had more than quantitative physics in mind when they pondered such things.

According to Greenblatt, the atomic nature of reality affects our innermost personal lives as well as our bodies:

> If you can hold on to and repeat to yourself the simplest fact of existence—atoms and void and nothing else, atoms and void and nothing else . . . your life will change. You will no longer fear Jove's wrath whenever you hear a peal of thunder, or suspect that someone has offended Apollo whenever there is an outbreak of influenza.

In this age of nuclear power and nanotechnology, it can be difficult to believe that our more detailed grasp of atoms is so new, but there are people alive today who were born before the scientific community as a whole was convinced that atoms are real. It was only in 1803 that the British chemist John Dalton proposed a formal atomic theory, and it was only in 1905 that Albert Einstein demonstrated the existence of molecules and atoms through a mathematical analysis of Brownian motion.

Today specialists routinely study individual atoms and their precise arrangements in molecular structures. But even for those of us who lack the requisite equipment, simply knowing for sure that atoms exist can be as life-changing as it was during the time of Lucretius.

On the size scale of elemental particles, everything is always in motion. Even in the calmest puddle or the hardest ice, atoms are shivering, not from cold but from heat. The universal thermal dance of matter is one of the least structured of all movements. It is as close to true randomness as the real world can get, so close to it in fact that mathematicians use equations based on it to drive their random-number generators.

The likeness of Brownian movement to true dancing is more than coincidental. In a paper titled "Collective Motion of Moshers at Heavy Metal Concerts," physicists at Cornell University studied closely packed dancers in videos from a rock concert. They found that disordered gas equations, when scaled to accommodate the two dimensions of a dance floor, nicely described the movements. "These findings," they wrote, "offer strong support for the analogy between mosh pits and gases." Published in the February 11, 2013, issue of *arXive.org*, the study was done in good fun, but the science behind it is sound enough to have predictive value in studying the dynamics of mob behavior.

Of course, a real dance is done with intent, and atoms lack the hearts, brains, and feet for that sort of thing. But the analogy describes the restlessness of atoms in ways that help to illustrate real-world phenomena such as temperature.

Normal conceptions of temperature and heat don't work as you might expect when you consider them on the atomic level. In that realm "heat" is a kind of energy that passes from one particle to another when they collide. The more heat, the more vigorously the particles bounce. In a way heat is much like the vital force that many of Brown's contemporaries believed in, but it applies to all matter. This background energy is present wherever atoms exist, and all atoms dance to it. Even when tightly bound in the lattice of a quartz crystal or a chunk of bone, atoms fidget in place like an audience feeling the beat from their seats during a rock concert.

Temperature, on the other hand, is an effect of heat. Both the temperature and the mercury rise within the glass tube of a thermometer when increasing heat speeds the jiggling of the mercury atoms, forcing them to spread more widely as they strike one another more violently. Similar processes occur on a crowded dance floor. People can pack tightly together when they stand still, but when the music starts and they begin to jostle one another, everyone naturally spreads out more. The temperature of an object, then, reflects the vigor of the heat-powered dance of particles within it, and the expansion of heat-stimulated gases explains why hot air rises and drives the dynamics of wind and weather.

The Cornell study on mosh pits found that the equations of temperature also apply to a dance floor, as well. Using computer simulations, the investigators modeled the loss of energy in Mobile Active Simulated Humanoids through collisions between a dense cluster of simulated MASHers and the bystanders that surrounded them. "Measurements *in silico* [dancers on a computer] show a radial temperature gradient is established," they wrote, "with a higher effective temperature at the core of the simulated mosh pit and a lower effective temperature at the edge."

To bring the heat-driven dance of atoms back to you in simpler terms, thermal motion on the atomic scale helps you to live and breathe. Because of it oxygen diffuses into your blood from your lungs, and signal molecules leap the narrow gaps between your nerve and muscle cells so rapidly that you can jerk your hand away from a hot surface before it burns too deeply. It is also the trembling of atoms and molecules that transmits the warmth of your touch. When you remove a glove to grasp someone's cold hand on a wintry day, the collisions from your more energetic atoms invigorate their more sluggish ones, while waves of diffusing ions telegraph the experience through your nervous system to your brain and, perhaps more metaphorically, to your heart too.

What role does hydrogen play in all this? For one thing, nearly two-thirds of your body consists of hydrogen-bearing water molecules. If you are a 150-pound adult, then most of you exists only because roughly 10 pounds of hydrogen atoms clung to 80 pounds of oxygen atoms and somehow made their way into your body in a manner that allows you to exist at this particular moment. You might be even happier that they did so after taking a closer look at what they do inside you.

If you want to understand how the vibrant atomic nature of things helps to explain why you are what you are, then it also helps to remember that a water molecule, in silhouette, looks like Mickey Mouse's head.

When two hydrogen atoms ride an oxygen atom, they take up positions on opposing sides of the same hemisphere. Because of this the molecule looks like a spherical mouse head with two rounded mouse ears. The positioning of the hydrogens produces many of the unique, life-sustaining properties of water, because it gives the opposite ends of the molecule slight ionic charges of opposite signs. In the world of water molecules, ears are slightly positive and chins are slightly negative. Why does this matter? Opposites attract, so the asymmetry makes water molecules tend to stick to one another, ear-to-chin, as well as to other objects that carry a charge. Although these "hydrogen bond" attractions are weak compared with the covalent bonding that holds a molecule together, their effects on you and the rest of the world are impressive.

You can feel the influence of hydrogen bonds on you when you sweat. The evaporation of sweat is the transformation of crowded molecules of liquid into flying specks of gas as heat from your skin goads them into more vigorous motion. The more hydrogen bonds the heat breaks, the more water molecules bounce free of their partners. These escapees disperse as vapor and haul body warmth into the air with them, leaving you cooler.

Water vapor can comprise as little as 0.1 percent of the air around you (in a desert) or as much as 4 percent or more (in a humid rain forest), but its individual molecules are too small to reflect visible light as clouds do. The steam that appears out of thin air a short distance from your lips on a chilly morning does so only because the cold slows the thermal dancers in your breath enough for hydrogen bonds to do their sticky work more effectively. They can then pull the water vapor molecules together into glistening droplets that become just large enough for you to see but still small enough to fly.

Although most of your body water enters you in food and drink, vaporized water molecules also bump into you on every breeze and rush into your airways with every breath. Some of these diffuse through the spongy interiors of your lungs and enter your bloodstream along with the oxygen that you inhale. Hydrogen bonds in your tears also pull airborne moisture into your eyes, and other water vapor molecules are, at this very moment, wriggling into microscopic crevices in your hair.

Speaking with someone who cuts hair for a living can also be enlightening on this last point. On my last visit to the person who cuts mine, she deepened my appreciation for the connections between my hair and gaseous water. When I asked Patty McDonough if she knew that water vapor from the air can seep into hair, she rolled her eyes as if I had just asked her if she knew what a pair of scissors looks like.

"Yes, of course," she said. "The shafts of your hair are scaly like a pine cone. A blow-dryer or hot weather can open the scales up and expose the insides to the air. When that happens, moisture soaks in more easily and it makes your hair look thicker."

Curly hair in particular has scales that don't overlap as tightly, so water vapor enters more quickly through the gaps. You can sometimes watch this kind of hair change before your eyes in humid weather. "I can tell right away if it's humid outside," Patty said, "because my customers' hair puffs up more than usual."

In the wilds of upstate New York where I live, a bad-hair day is no big deal; at least not for most of the people I hang out with. We regularly abuse ours with sweat, hats, and bug dope. But the air-hair connection is apparently important enough to some people that a Web site titled Hair Forecast caters to their needs by monitoring atmospheric conditions in hundreds of American cities and posting daily "hair index" ratings. Simply type in the name or zip code of that city you're about to visit, and you can arrive prepared for a dry, sunny "10" or a damp, frustrating "1" in which you can expect, as the Web site puts it, "major frizz."

This exchange of water molecules between you and the atmosphere means that you are more or less continuous with your surroundings. And although water comes and goes easily in vapor form, the atoms that comprise it also build more durable structures into the fabric of your body. As your hair absorbs and releases moisture, the proteins in those filaments also trade hydrogen atoms with the passing water molecules. And not only does your hair share hydrogen with the air around you: You also share hair atoms in this manner with everyone else who happens to be in the room with you.

The continuous reshuffling of hydrogen atoms among molecules happens because the covalent bonds that tie them to oxygen atoms are impermanent despite their strength. Water is a composite collection of atoms that can lock tightly together like nuts and bolts but can also disconnect and rearrange themselves into the same number of identical molecules in different combinations. Although it is fun to imagine a prehistoric ancestor slurping the same molecules that fill your drinking glass, water's innate tendency to share its atoms with neighbors makes this unlikely. The hydrogen and oxygen atoms in your body may be billions of years old, but the molecules in which they travel are temporary.

It is easy to imagine that atoms carry with them features they once helped to create in previous molecular forms just as, for example, a lunchbox may carry the lingering odor of food that it once contained. But although a water molecule carries no memory of its former hosts, people's difficulty with believing this can have important consequences for the management of water resources, particularly when shortages in arid regions raise the possibility of recycling water from human wastes. Despite the hard sell, city officials in Australia, Singapore, and elsewhere have managed to persuade their citizens that water that has been thoroughly treated in "toilet to tap" systems is actually purer than much of what normally comes out of a faucet or commercial bottle. Such reclamation currently supplies a third of the drinking water for the city of Windhoek, Namibia, and astronauts routinely drink water extracted from their own urine with no adverse health effects or aftertaste. But it can be so difficult to think of your food and drink in terms of molecules and atoms that initiatives to recycle waste water are often rejected by voters, even where shortages approach crisis levels. In a blog post from the Earth Institute of Columbia University, the environmental engineer Shane Snyder is quoted as saying that, fortunately, most people do understand the need for this kind of reuse as long as they trust the utility responsible for it. Virtually all the water molecules on this planet's surface are recycled, and as Snyder explained, "We're going to drink recycled water one way or another, whether it comes from downstream flow or groundwater. I strongly believe we should to do it through engineered systems where we can actively control the process."

About one-tenth of the hydrogens in your hair are hitchhikers that camped in your keratin within a few hours of your arrival in a new location, reaching full equilibrium with the water vapor in your surroundings within three or four days. But most of your structural hydrogen, which is more firmly embedded in your proteins and other biomolecules, enters your body with food and drink instead. We know this because scientists can now track the movements of atoms with the help of stable isotopes, odd varieties of atoms that differ slightly from the norm but not enough to merit a new place on the periodic table of elements. As drifting smoke and smells betray invisible currents in the atmosphere, isotopes can allow scientists to trace the flow of atoms into, through, and out of your body.

A naturally occurring isotope of hydrogen, called deuterium, formed in relatively modest amounts along with regular hydrogen shortly after the Big Bang. It differs from hydrogen only in the presence of a neutron alongside the usual proton in its nucleus, and you could drink deuterium-based water without noticing anything out of the ordinary. The extra mass of this isotope, however, makes it useful as a sort of atomic identification tag. When water evaporates from a lake, for example, the normal molecules fly out of the crowd more easily than the ones with overweight deuterium "ears." In a process much like winnowing wheat from chaff by tossing the mixture on a windy day, a warm or dry setting tends to enrich lakes and groundwater with isotopically heavier water. This sorting effect can make different water samples distinct enough from one another for scientists to deduce the climate, elevation, or latitude they came from. And because you and other living things consume deuterium-bearing water, those telltale atomic signatures also appear in you.

Thousands of isotopic analyses have been conducted on water sources around the world, particularly in North America and Europe, and detailed maps can now help to match the deuterium contents of human hair, fingernails, and bones to specific regions. You can even type your latitude and longitude into an "Online Isotopes in Precipitation Calculator" maintained by Waterisotopes.org and find the average deuterium:hydrogen ratio in your local precipitation. Such maps are called "isoscapes," a casual blending of "isotope" and "landscape."

In a paper published in the *Proceedings of the National Academy of Sciences*, the University of Utah ecologist James Ehleringer and his colleagues analyzed slight variations in the amount of deuterium relative to normal hydrogen in human hair and found that the atoms of groundwater become integral parts of people who drink that water whether they get it from a tap, a cup of coffee, or the milk of local cows. This transfer of atoms from groundwater to body water to more solid body parts is slower than air exchange, taking about a week for hair follicles to deposit local hydrogen into the roots of growing hair shafts, for example, but it is more permanent once it happens.

The abundance of deuterium in the solid parts of your body reflects the isotopic composition of the water you consume, which in turn reflects the climatic conditions where your water originated. In the aforementioned article, the deuterium isoscape of the United States is laid out in color, with brilliant orange-reds indicating higher values in balmy Texas grading into deep blues for lower values in the cooler, moister northwestern states. After analyzing hair clippings from the floors of barber shops across the country, Ehleringer's team found that not only do the atoms of local water appear in the bodies of local residents—the isotopic balances of the water do so, too. Citizens of Texas, for example, tend to produce measurably heavier hair than do people from the northern Rockies.

The more localized your water sources are, the more your isotopic composition matches that of your location, but if you drink a lot of commercial bottled water you may be more closely linked to distant places instead. A Florida resident who consumes spring water that was bottled in Maine, for instance, carries less deuterium in his or her body than neighbors who drink local tap water, even though they both shop at the same supermarket chain.

In a study led by Gabriel Bowen of the University of Utah, researchers analyzed the isotopic signatures of bottled waters used by American soldiers in a military base in Baghdad, Iraq. The diverse selection displayed a wide range of deuterium fractions, which reflected their imported status. The lightest water came from European sources, and the heaviest came from Saudi Arabia, as would be expected from the differences between wet-cool and hot-dry settings. Most importantly, however, the samples contained significantly less deuterium than the local precipitation, confirming that none of the water was bottled in Baghdad. Although the deuterium in water has no direct health effects on those who drink it, using it as a tracer in this manner can help to identify potentially contaminated domestic water that is sometimes deceptively sold in recycled bottles.

Physical linkages to local water are also clearly revealed in hair, which represents a sort of strip-chart recorder of its atomic sources as it grows. In the Ehleringer study a man who had recently moved to Utah from China was found to carry an isotopic record of his travels on his head. By analyzing deuterium concentrations at regular intervals along a single filament, the researchers could reconstruct when the man had arrived at his new home. Within a month of consuming the lighter water of Salt Lake City, the base of his hair shaft contained much less deuterium than the older portion that grew in Beijing.

So persistent are the atomic connections between people and their environments that archaeologists use them to study ancient history. In one study published in the *Journal of Archaeological Science*, a filament from the head of a centuries-old mummy revealed clues about the last months of an Inca child's life. High deuterium content in the hair as a whole was interpreted to mean that the boy spent most of his final year drinking relatively warm waters at an elevation of about a mile, far below the frigid Andean peak where his body was found. A slight decrease in the middle portion of the filament identified protein that formed in winter, when evaporation rates and deuterium fractions in local waters were lower. High concentrations of deuterium close to the scalp showed that death came during summer, and that the boy spent less than a week at high altitude before he was sacrificed.

If you, too, were to become a dehydrated mummy, then hydrogen and oxygen atoms would still be common in your leathery remains, but they would be dispersed in solid form rather than in water molecules. A molecule of your blood sugar, for instance, contains a dozen hydrogen atoms and six oxygen atoms arranged around a ring of carbon, and the sturdy fibers of your muscles are also well stocked with these same elements. Most of your structural hydrogens can be traced back to your food, but between a quarter and a third of them were extracted from water that you drank during the past year.

In the larger atomic picture of life, however, food hydrogens ultimately come from water. The hydrogen atoms in meat and milk, for example, were recycled from drinking water and plant tissues that animals consumed. Most of that vegetable matter, in turn, was fashioned from carbon dioxide and water by the plants, and it is no coincidence that the metabolic breakdown of your blood sugar yields equal measures of CO_2 and H_2O. The dominant atoms in a pound of flesh or fava beans can, with the help of breath oxygen, be rearranged into an equivalent mass of carbonic gas and mist.

If you could probe beneath the surface of your body on the atomic scale, you might be surprised to see how the motions of water molecules keep you alive. A more in-depth tour of your atomic self therefore seems in order.

Imagine shrinking down a thousandfold in order to follow a gulp of tap water into a human digestive tract and out again on a damp breath. In this thought experiment you are still far too large to follow the water molecules into the bloodstream, but already the liquid surrounding you feels very different from your typical beverage. The water molecules cling to one another with hydrogen bonds, so it is more difficult for you to force them apart. Now the water feels like syrup, thick and sticky.

Reduce your size another thousandfold, and you are still ten thousand times larger than the water molecules. As a Brownian particle yourself, you are now more violently buffeted by them, and they never let up. Each one is moving faster than a jet airliner, so you're lucky that they aren't heavier. But even now you're still too large to follow the water on its journey.

Drop a final ten thousandfold step to the approximate size of a water molecule, and kick your imagination into overdrive. Unfortunately the molecular collisions become intolerable now that the impactors are the same size as you. Use whatever sleight of mind you wish to deal with these problems, and follow the water.

Originally the gulp of liquid seemed to be a more or less coherent substance moving in a single direction through your gut. But in this chaotic realm of tumbling, twitching particles, the most obvious motions are driven by heat and are therefore mostly random. Like your molecular neighbors, you lunge a short distance, career off something, then ricochet off something else, over and over and over. How are you going to get anywhere on purpose or on time?

If you need to move large amounts of matter quickly in a controlled fashion, your body can use muscles and skeletal levers to do the work, but it takes energy to do it. For example, the muscular contractions of your stomach help to move and process your meals, but those muscle cells also have to be fed in return. Such bulk movement as this is more of a costly, organized migration than a dance.

On the other hand, you can also have things delivered for free, but only over short distances. To do so you can harness the jiggling of atoms and molecules to push things around, and it won't cost you a dime's worth of energy. The interiors of your cells churn with Brownian motion, which helps to move resources and wastes to their destinations. The buffeting pressure of your dancing water molecules also helps genes and proteins to fold and hold their proper shapes. Even pathogenic viruses, which lack limbs to swim with, rely on the thermal movements of your water molecules to jostle them into contact with your cells, which they can then latch onto and infect.

The motion is erratic, but it is so blindingly rapid that even a million false moves can still lead to a desirable goal very quickly as long as the route is short. Think of it as trying to make your way out of a crowded room with your eyes closed—give yourself enough time and sooner or later you'll find the door. On the atomic scale "sooner or later" can play out within a tiny fraction of a second.

This restless dance of atoms and molecules itself makes it advantageous for your body to be comprised of cells rather than some homogenous material. As a cellular being you use both methods of transport at once but in different situations. Energy-intensive bulk flow through organs and vessels carries large amounts of matter most of the way to wherever they are needed, as trains deliver goods to distribution centers. From there, cost-free thermal movements can take over for the brief random drift into the interiors of your cells. The trip must be short, however, because the sheer mindlessness of it makes the duration of a journey increase with the square of the distance.

According to an online calculator at PhysiologyWeb.com, it takes an oxygen molecule only a few milliseconds to diffuse into the center of a red blood cell as it squeezes through a capillary in your lungs. But for it to diffuse over greater distances would take much longer because the molecule would keep wandering to and fro rather than aiming for the target. To traverse the length of your thumbnail, for instance, would take several hours, and to reach the base of your thumb could take a week. It is not diffusion but bulk currents that bring you the scent of perfume from someone standing nearby because it could take years for scent molecules to cross a room by diffusion alone. Nonetheless the short-distance Brownian migrations we call "osmosis" (for water) and "diffusion" (for other molecules) drive food, gases, and fluids into or out of every cell in your body, and they do a fine job of it at little or no energetic cost to you. A signal molecule, for example, can jitter from one nerve cell to another in about three millionths of a second and thus help you to respond rapidly to your environment. This combination of directed and random motions underlies the energy economics of all life on Earth.

As you continue this thought experiment, prepare now to ride a river of blood. This will be the first step leading from the open-ended tunnel of the digestive tract into the true interior of the body.

The inside of any cell is mostly water, so it's a wild trip through the cells of the intestinal lining. Everything is moving, and the vigorous dancing of the diverse molecules that you encounter here makes it easy to forget that life as we know it exists only on larger size scales.

What appears to be a lumpy cable car slides along a taut strand of protein that connects one corner of the cell to another. This is a kinesin, a transport machine that runs on Brownian motion. As restless water molecules bash the kinesin, a ratchet mechanism prevents the thing from moving backward and produces a net forward slide along the cable. In this manner your own heat-driven kinesins help to supply the inner workings of cells with raw materials that come in through your mouth.

Nearby, a tightly wrapped knot of protein convulses among the water molecules that swarm around it. It is a digestive enzyme, so you had better keep your distance. This one specializes in chopping sugar molecules into pieces, some of which may later leak from the lungs as carbon dioxide. The shape of the enzyme allows it to clutch a sugar and then bend in such a way as to break its victim in two. As with all such molecules, the shape of this sugar chopper is not only a feature of the enzyme itself but also of the water that surrounds it. Slight charges on the enzyme's irregular surface attract water molecules that press against it. This makes some sections warp outward and forces others inward, sculpting features that help the enzyme to do its job of butchering food for you.

The thermal frenzy of water molecules is so pervasive that all cells depend on it to help drive their metabolisms. Trapped inside each mitochondrion are hordes of hydrogen nuclei that have been pumped into the center of the blob. They seethe as if eager to escape, but a specialized membrane confines them. Tiny holes in the membrane let them stream out through molecular machines that package chemical energy for use elsewhere. In a warm-blooded creature such as yourself, mitochondria can also produce heat for a Brownian thermostat that coordinates the rates at which a cell's contents move and interact.

You will now need to travel more quickly and purposefully through the bloodstream in order to reach the lungs. A fleshy valve opens, and you're sucked into the heart. A firm squeeze, and you tumble out of the right ventricle. A moment later you enter a maze of lung capillaries, where diffusion drives you into a tiny air sac. A tightening of the chest muscles, and out you go in a rush of warm, humid breath.

If you were to continue to follow your watery companions through the atmosphere, you might eventually watch them condense and fall from the sky anywhere in the world. It could then be only a matter of days before they re-enter the realm of the living. Your body is only one of many homes that your hydrogen and oxygen atoms have occupied throughout their long histories, and the other creatures that share this planet with you may also share these very same atoms in much the same manner.

Here, then, are three examples of that mutual atomic heritage, two of them from recent times and one from the distant past.

Hydrogen-bearing keratin proteins like those in your hair can also build beaks, claws, hooves, and feathers. Because of this, wildlife biologists can use deuterium to determine where animals travel and how their behavior changes with the seasons, because the atomic fingerprints of drinking water appear in their keratin.

One such study, published in *Oecologia* in 2000, uncovered a remarkably close atomic relationship between saguaro cacti and white-winged doves in Arizona. Samples of body water collected from doves showed that their deuterium balances varied from month to month. Concentrations resembled those in local rainwater from early May until mid-June, but then rose abruptly. The bodies of the doves, it seems, were mirroring the annual cycle of flowering and fruiting in the saguaros.

Saguaro nectar contains more deuterium, ounce for ounce, than groundwater does because the plants lose a lot of moisture to evaporation in the dry desert air. When cactus flowers bloomed after seasonal rains arrived, the doves began to drink more nectar than groundwater. Within days the body water of the nectar-sipping birds was enriched, too. Later on, when the cactus fruits ripened, the birds switched to eating fruit pulp and seeds, and that dietary shift boosted the deuterium contents of the birds' body fluids even further.

On a practical level this research showed that the white-winged doves of Arizona rely on cacti for most of their nutrition, but it also showed how literally their habitat makes them what they are. The same, of course, is true of you. Even if you live in a major city you still construct your body from the same earthly reservoir of atoms that the doves and cacti of Arizona do.

In a study reminiscent of Ehleringer's work on human hair, the geochemist Thure Cerling and an international group of colleagues used isotopes to trace the atomic connections between elephants and their surroundings in northern Kenya. Appearing in the *Proceedings of the National Academy of Sciences* under the title "History of Animals using Isotope Records (HAIR)," the results of that study showed that the tail hairs of elephants are built in part from raindrops and river water.

Four members of a single family, dubbed the Royals, were equipped with radio collars for tracking, and tail hairs were removed while the animals were immobilized for fitting and the replacement of batteries. The isotopic composition of local river water was analyzed along with that of the hairs, and precipitation was monitored at a nearby weather station. When all the records were compared after six years, they revealed a remarkable pattern.

During wet seasons the deuterium fraction of water in the Ewaso Ng'iro River decreased sharply, as cloudy weather and rain slowed evaporation and diluted groundwater. Almost as rapidly, the deuterium content of the new keratin in the tails of the Royals dropped in response, as the elephants continued to pack the hydrogen atoms of local rivers into their hair proteins. Growing by as much as a quarter of an inch per week, the Royal hairs preserved a molecular record of their atomic connections to the world.

Because oxygen is also a component of water, the links between water and animals can be tracked with oxygen isotopes as well as with deuterium, and Daniel Fisher, a paleontologist at the University of Michigan, uses them to study large mammals that died thousands of years ago. By measuring the ratios of heavy oxygen-18 to normal oxygen-16 in ancient tusk ivory, he has helped to revolutionize the study of mastodons that wandered North America shortly after the last Ice Age.

After cutting and polishing slices of well-preserved tusks that were unearthed in the Great Lakes states, Fisher noticed narrow bands that resembled the annual rings in trees. As with today's elephants, the tusks of mastodons were elongated incisor teeth that accumulated in distinct layers throughout the animal's lifetime just as the tusks of elephants do today. Fisher knew that the dental calcium and phosphorus in ivory is sprinkled with oxygen atoms, and he therefore surmised that oxygen from lake water, river water, or even snow lay in those old tusks.

Powdering and analyzing sequential samples for their oxygen content, Fisher and his colleagues measured the relative abundances of isotopes in the concentric ivory layers. Like deuterium, heavy oxygen-18 also tends to become more abundant in water under warmer conditions, and oscillating oxygen isotope ratios in the rings suggested that the temperature of the animals' drinking water rose and fell cyclically. In other words these mastodons were so atomically tuned to the temperatures of local waters that they carried seasonal climate signals in their tusks.

By confirming that the bands were annual growth rings, these results revealed not only the ages of the animals but also their sex. Some of the tusks displayed more variable growth than others: The youngest rings at the center were fairly uniform, but between the ages of nine and twelve, the bands became thinner. From then on their thicknesses changed every three or four years, as would be expected if the animals were devoting nutritional resources to something other than their own growth. The more variable tusks clearly belonged to reproductively mature females—mastodon moms.

A sudden drop in ivory production for one female from western New York represented her first pregnancy at age nine to ten, and the thinnest bands after that represented the first two years of nursing a calf. This mother carried the marks of half a dozen pregnancies and weanings in her tusks before her death at age thirty-four. Judging from the water temperatures indicated by the oxygen isotopes, she probably gave birth in spring, and repeated zones of thinner rings also showed that she continued to nurse her babies for as long as four years. As a youngster became more independent and no longer pulled atoms away from its mother, the mother's tusk rings thickened again.

Each mother mastodon in Fisher's study physically diminished herself on behalf of her children, and her nutritional sacrifices left indelible atomic imprints on her remains, as did the passage of the seasons, the plants she browsed on, and the water she drank. Now, thousands of years later, we can read that story engraved in ivory long after the last of the mastodon moms melted back into the global pool of elements that now sustains you and me.

When viewed from space, our home planet is mostly blue, white, and green, and all but a small fraction of it owes its color to water molecules. Even on this enormous scale, you can see signs of the atomic vibrations that help to determine the distribution and nature of life on Earth.

The blue areas represent water molecules that are jiggling rapidly enough to flow but slowly enough for their hydrogen bonds to hold them down and out of the atmosphere. The creamy clouds are cool droplets whose sticky hydrogen bonds pull vapor molecules in from faster thermal dances in the air, and they move on winds that are driven by the expansion and contraction of huge parcels of air whose molecules tremble differently at different temperatures. The snow and ice represent chilled, slow-dancing molecules whose hydrogen bonds freeze them into porous crystal lattices that—fortunately for fish—can float rather than sinking and filling lakes and oceans from bottom to top with petrified water every winter.

The green patches on land also represent water in action. According to a recent paper in *Nature* by hydrologist Scott Jasechko and colleagues, plants pump approximately 15,000 cubic miles of groundwater into the atmosphere by evapotranspiration every year, nearly enough to drain all of the American Great Lakes every four months and more than the annual flow of all the world's rivers combined. The exhalations of forests, fields, and grasslands therefore drive the largest movements of fresh water on Earth.

But what ultimately links air, oceans, and plants most tightly is the hydrogen-fusing sun. It is easy to ignore it as we go about our business. But what if it weren't there, even for just a few seconds? Everything that is not artificially lit would vanish into darkness—we forget that we can only see a landscape, a city street, or even the moon because sunlight reflects away from it and then strikes our eyes. Without sunlight, the noon sky would become a transparent window on the galaxy, and terms like "day" and "night" would become meaningless. All outdoor photosynthesis would cease, and if the shutdown persisted long enough the oxygen concentration of the atmosphere would begin to drop more rapidly than it already does.

But long before our oxygen supply could run out, we would either die of thirst or freeze to death because solar heat drives the thermal motions of molecules on Earth at a higher pace than a sunless world would permit. Temperatures on the frigid dark side of the moon can approach minus 300°F, and even though the heat-trapping greenhouse gases of our atmosphere insulate us through the darkness of night here on Earth, most of the energy in that airy blanket came from the sun. Without our companion star beside us normal weather systems would shut down, leaving the rainless continents parched as the oceans slowly froze.

Astronomers know well what it can be like on a planet whose thermostat is set higher or lower than ours. Venus has more greenhouse gases and flies closer to the sun than we do, and its water molecules vibrate too rapidly for hydrogen bonds to hold them in liquid form. On Mars the opposite situation prevails. With much less solar heat to goad its molecules, Martian water is mostly solid, though single molecules do shake loose to fly off and re-attach to ice crystals elsewhere when the distant sun warms them enough.

On Saturn's icy moon, Titan, water is a stone from which landscapes are constructed. Titan is studded with rivers and lakes of liquid methane and ethane whose molecules dance slowly enough in the brutally cold atmosphere to condense from hydrocarbon clouds. Imaginary residents of Titan might wear water ice in their jewelry as we wear sapphires, and they would probably gasp at the thought of living on a planet such as ours that is mostly submerged under molten blue oceans of water lava.

Our home planet supports life as we know it because it has lots of liquid water. This, in turn, is because we circle the sun at just the right distance for water molecules to shift among gas, liquid, and solid phases in different altitudes, latitudes, and seasons. The thermal motions of Earth's surface molecules are tuned to such a narrow and serendipitous range of temperatures that a difference of a single degree can turn rain into snow or a rippling lake into a rigid, frosty plain.

The dance of atoms that early Greco-Romans uncovered through sheer deductive reasoning can now be studied and savored in delicious detail, as can the most fundamental elements of life. Every atom in your body exists only because hydrogen appeared in the universe billions of years ago, and the water molecules that surround and suffuse you owe their distinctive properties as well as their very existence to primordial hydrogen. Anything that has to do with water, from the green tissues of a leaf to the moist bag of cells that is your body, is here because hydrogen atoms ride oxygen atoms all over the world and tremble just so in the warmth of a hydrogen-fueled sun.

Hydrogen, given enough time, does indeed become people. How utterly amazing it is how miraculous you are, and how wonderful to be able to recognize and appreciate it, too.

3

Blood Iron

*. . . how very intimately the knowledge of properties and
uses of iron is connected with human civilization.*
— George Fownes

*Remember the terrible song of stars—you knew it once,
before you were born.*   — John Daniel

On a clear morning during the summer of AD 1054, a Chinese astronomer named Yang Weide noticed something sparkling in the
eastern sky that he had never seen before. Sending an urgent message
that is still preserved in the official records of the Song Dynasty, he
reported, "I humbly observe that a guest star has appeared." He could
not have known it at the time, but Yang Weide was describing one
of the most violent phenomena in the universe. He was also recording the distant, explosive birth of atoms identical to those that ran
in his veins, comprised his scientific instruments, and built the very
Earth upon which he stood.

According to astronomical documents the star was initially reddish but it turned pale yellow as it moved higher above the horizon,
much as the sun does when shining through dusty air in that region
of China. Yang surely recognized the political value of such details.
Yellow was the imperial color at the time. "I respectfully submit,"
Yang wrote in his official interpretation of the event, ". . . an abun-
dantly enlightened one is ruler, and the state has great worthies in
office."

In his analysis of the Chinese literature, the linguist David Pankenier noted that Yang, as a senior administrator in the emperor's Bu-

reau of Astronomy, had an obvious incentive to flatter the throne by putting a positive spin on what was happening in the sky. But then as now, political intrigue was as much a part of human life as were physics and chemistry. A palace official named Zhao Bian, who was involved with the impeachment of the grand chancellor at the time, put a darker spin on the star visit. His writings referred to calamities that had accompanied a previous "guest star" in AD 1006, and he concluded that this latest appearance portended troubles from banditry to earthquakes, all of which reflected badly on the emperor's administration.

The Song documents noted that the star could be seen in daylight for several weeks, then became visible only at night and seemed to vanish nearly two years later. The Bureau of Astronomy reported, "A guest star had appeared at dawn, in the direction of the east, under the watch of Tianguan. Now it has disappeared." On Chinese sky charts Tianguan lay close to the Three Stars constellation, also known as Orion, and the historical records of the guest's position allow today's astronomers to tie it to what is now its diffuse remnant, the Crab Nebula which lies a short stretch upslope from Orion's tilted belt.

Meanwhile, far to the west, a scholar in what is now Iraq recorded the same event. The philosopher-physician Ibn Butlan wrote that "one of the well-known epidemics of our own time has occurred when the spectacular star appeared. . . . In the autumn of that year, fourteen thousand people were buried . . . in Constantinople." In western Asia as in the East, prominent scientists were more open to the idea of causal links between astronomical and human events than they generally are today. "As this spectacular star appeared," Ibn Butlan continued, it also "caused the epidemic to break out in Old Cairo."

Although we might now dispute such direct causal relationships among stars, plagues, and politics, Ibn Butlan and Yang Weide were correct that hidden ties can join seemingly disparate aspects of existence. And more recent investigations have also shown that stars really do influence disease and human society, albeit indirectly through the metallic atoms that emerge from them.

It wasn't until nearly nine hundred years later that Albert Einstein helped to make more demonstrably sound connections between space-explosions and our bodies. His famous formula, $E = mc^2$, equates energy with matter and helps to explain how dying stars create the elements of life. "It is a glorious feeling," he once wrote to a friend, "to recognize the unification of a complex of phenomena that appear . . . as completely separate things."

According to the biographer Walter Isaacson, Einstein's fascination with mysterious unseen forces dated back to early childhood. One day when he was four or five years old, his father gave him a compass to comfort him while he lay sick in bed. The wonder of watching the compass needle move as if guided by an invisible hand followed him through his later explorations of magnetism and gravity. Years afterward he explained, "I can still remember—or at least believe I can remember—that this experience made a deep and lasting impression on me. Something deeply hidden had to be behind things."

But even Einstein did not have the information necessary to trace the finer points of the atomic connections between a Chinese guest star, the germs that sickened him as a child, and the strangeness of a compass needle. Like Yang Weide and Ibn Butlan before him, he died before more recent findings revealed such things. Our current explosion of knowledge, triggered by a long line of technological innovations, could only have happened with the help of one element that all these disparate people and phenomena have in common.

Iron, the star killer, is arguably the most destructive element in the universe and, at the same time, a master key to human existence. Your body uses it as a tool to harvest oxygen from the air as well as a weapon against microbial invaders, and its use on larger scales has helped to create and destroy civilizations. Iron shields you from the death throes of distant stars while also hurling meteoric missiles at you from the depths of space, and its story helps to reveal a deep physical kinship that you share with the rest of the cosmos.

Iron occupies the twenty-sixth place on the periodic table of elements, and is the sixth most abundant atom in the universe. It is heavier than most of the other atoms of your body, carrying within it twenty-six protons and thirty neutrons, and the large nucleus anchors a thick cloud of electrons with which it can make multiple bonds to other atoms. These features help to explain iron's role in the destruction of stars and its remarkable properties here on Earth.

Atoms are fussy about the arrangements of the electrons surrounding them, much like people who are obsessed with their attire. In the company of other atoms, iron often tries to donate or share some of the electrons in its outermost shell or to steal more of them from neighbors. One of its favorite trading partners is oxygen. Expose iron to air for long enough, and it will corrode into rusty oxides. But if you embed a single iron atom in an oxygen-toting hemoglobin molecule in your blood, oxygen's attraction to iron becomes more beneficial to you.

Another possible companion is carbon, which in standard concentrations of 2 percent or less hardens iron into steel. Clutch one end of an iron rod tightly enough, use a machine to pull on it strongly enough, and you can draw it like taffy through a reducing hole in a sturdy steel die, then another and another until it thins down into a string. If the strand also contains the right amount of carbon and has been heated just so, you can then stretch it tightly across the bridge of a violin to pluck and bow sweet music from it without snapping it. This strong, ductile nature arises from the ability of iron atoms to act like ball bearings with attractions that bind them together but also allow them to slither around one another when pushed or pulled. These tenacious but malleable bonds arise from interconnected clouds of electrons.

Pound an iron rod on an anvil with a heavy hammer, and it will flatten into a slim blade as the atoms within the metal spread sideways without breaking their grips on one another. Heat the rod in hot coals first, and the more vigorously dancing atoms will slide more readily into new shapes. Heat it to 2800°F in a properly stoked furnace, and the trembling, glowing atoms will flow into whatever container awaits them, as early ironworkers discovered to their pleasure many centuries ago.

If you expose an iron rod to an electrical current it will conduct it like water through a hose. This is because the metal's loosely bound "sea" of shared electrons flows so freely from atom to atom that a lightning rod can drain a thunderbolt into the ground. And if you expose the rod instead to a strong magnetic field, it can thenceforth attract the questing tip of a compass needle. The mobile electrons in the outer shell of an iron atom not only orbit the nucleus and occasionally dance off to visit neighbors: They also twirl as they go. The call of a magnetic field is like music to their ears, and they align their spins to match it. Many other substances do this as well, but iron can also remember the tune even after the music ends.

A magnetized needle generates its own field through the orientation of spinning electrons in closely packed atomic clusters that are roughly the size of a human cell. On that tiny scale they form a mosaic, each clump with its own spin direction. When a consensus majority of aligned clusters generates a net magnetic field, the needle can potentially move to align its own field with that of the earth's iron-rich core, which manipulates it through thousands of miles of overlying rock and magma. A sharp blow from a hammer, however, can shock the needle's electrons back into random alignments again, erasing the collective attraction to their siblings down below.

About two thousand years ago, early Chinese inventors hung bits of naturally magnetized iron ore, now called lodestone ("leading stone"), or magnetite, from strings that allowed them to point out the poles. Later "south-pointer" compasses of the Han Dynasty resembled delicate iron ladles resting on plates of polished metal upon which their slim handles could swing smoothly toward the south.

Over the centuries magnets such as these were used for geomancy, the enlistment of mystical forces to locate gems and other hidden objects, and for the alignment of buildings with earthly and heavenly features. During the Song Dynasty, Chinese navigators also used magnetic iron in military maneuvers and for trade across the Indian Ocean. Some of those compasses were iron needles that were stroked with magnetite before being placed on the surface of a dish of water or suspended from a silken thread. Sailors and traders of the time helped to spread compass technology throughout Eurasia, thereby opening the world's oceans to trade and contributing to the eventual arrival of European explorers in the Americas.

Nowadays we use magnetic metals for all sorts of things from computers and wind turbines to the engines of hybrid vehicles, and a new class of supermagnets now enlists unusually massive rare earth atoms such as neodymium to help unify and stabilize the alignments of iron electrons. This allows magnets to be built smaller, which makes them well suited for electronic devices such as iPods and earbuds as well as larger machines. The electric motor and batteries of a Toyota Prius, for example, contain about two pounds of neodymium.

Today China once again plays a major role in the spread of magnetic technology, this time as the primary global provider of rare earth elements. When officials recently reduced exports of these vital resources, world markets reacted with price spikes and a vigorous search for new deposits to exploit. Fortunately for non-Chinese consumers, neodymium is not as rare as its rare earth classification implies, and alternative sources in Australia, the United States, and elsewhere are under development at the time of this writing.

Pound for pound, neodymium magnets are many times stronger than your average fridge clinger. These are among the most powerful magnets on the market, capable of sucking the iron from a well-soaked sample of iron-fortified cereal and lifting a dollar bill free of a tabletop by the iron particles embedded in the ink. They also attract one another as well as other iron objects with surprising speed and vigor, which makes them potentially dangerous to handle. A blog post written in 2009 by the science writer Frank Swain, titled "How to Remove a Finger with Two Super-magnets," described—and illustrated with horrifying photos—what happened to one person's finger when it was caught between a pair of neodymium bar magnets as they slammed together from a foot and a half away.

Despite the risks, I recently bought a neodymium magnet of my own, for purely scientific purposes, of course. My goal was to hunt for an unusual class of iron. The metal in steel kitchen utensils and other such objects normally comes from two main ore minerals: dark, metallic-looking magnetite and reddish hematite. Most of the iron mined in the United States for the manufacturing, transportation, and construction industries comes from vast deposits of these ores in the Great Lakes region. Most of the iron in your body, however, can instead be traced to dispersed magnetite and hematite particles in soils and, to a lesser extent, to pyroxenes, micas, and other minerals in granite and basalt. But not all of it. Some of it comes from outer space.

People have long collected and marveled at iron that falls from the sky. Indigenous hunters shaped knives and harpoon points from a thirty-four-ton chunk that struck the Greenland ice cap thousands of years ago, and the mummy of the pharaoh Tutankhamun wore a dagger with a meteoric nickel-iron blade. Pieces large enough to turn into tools, however, are uncommon, and when a meteor as heavy as a hundred locomotives exploded over the Russian countryside in February 2013, it set off a "gold rush" of meteorite hunters who scoured the newly fallen snow for salable fragments and souvenirs. But extremely tiny meteorites are easier to find, especially if you have a neodymium magnet handy.

Careful to keep the magnet clear of the credit cards in my wallet, I swept it over a patch of grit where the steep roof of the building next to mine on the campus of Paul Smith's College sheds its rainwater. Dark pellets sprang up from the ground and smacked audibly into the shiny surface of the magnet. Peering at my catch under a microscope, I found two tiny metallic spherules glistening amid the irregular sand grains. These were frozen droplets of formerly molten metal that had sprayed the air overhead during the fiery deaths of meteors.

Lying amid ragged chunks of Adirondack magnetite whose atoms emerged from the same star-forges before the formation of the solar system, the little space spherules were not really aliens but long-lost relatives. Only by the most unlikely of navigational accidents were they now having their first family reunions in more than four billion years.

When a meteor blazes and seemingly vanishes, its atoms don't really disappear but remain aloft in more dispersed forms. So much meteoric metal disintegrates in the atmosphere that a mist of iron atoms several miles thick surrounds the earth at an altitude of about sixty miles. By some estimates as much as one hundred thousand metric tons of space dust settles quietly to the ground, the ocean surface, and the roof of your residence every year, including not only iron but also silicate minerals from asteroids, comets, and even occasional splash-ups from Mars. Many of these dissolve into groundwater and seawater and enter your body in atomic form through food chains, along with their terrestrial kin. As a result the fleeting streak of a meteor, a humble pinch of sand, a pharaoh's sacred dagger, and a wavering compass needle all share a common heritage with you.

Cosmologists propose several dramatic origin-stories for our personal elements, apart from the hydrogen which condensed shortly after the Big Bang, and all of them involve the destruction of stars. In a Type Ia supernova, for instance, one star steals matter from a nearby neighbor, then bursts after overeating. In a Type II "core-collapse" supernova, a star several times larger than the sun exhausts its fuel supply, falls in on itself, and then explodes. And in yet another scenario, an unusually massive star produces matter and antimatter particles that annihilate one another and their host in a "pair-instability supernova." Such cataclysms can now be seen directly or in hindsight with powerful telescopes on spacecraft such as Hubble and Spitzer, as well as from observatories on the ground.

The presence of iron in your body shows that your atomic ancestry can be traced back to exploding stars, and the supernovas and nebulas that are visible today can show you what those earlier cradles of creation looked like. One of them, the Crab Nebula, has much in common with the turbulent cloud of gas and dust that gave rise to our solar system. Through a telescope it may resemble a nondescript dab of mist, but it is in fact a huge storm of debris trillions of miles wide that once was a star eight to sixteen times more massive than the sun. Although the core-collapse explosion that produced it occurred more than seven millennia ago, the nebula we see now is still burning hot and expanding at roughly a thousand miles per second. Spectral analyses of the luminous knots, filaments, and fogs within it reveal that much of it consists of hydrogen along with iron and other elements. And with new information from the coldest place on Earth, we can now support historical links between the Crab and one of the hottest spots in the galaxy, the dying guest star whose light finally reached Earth in AD 1054 after crossing 6,500 light-years of space.

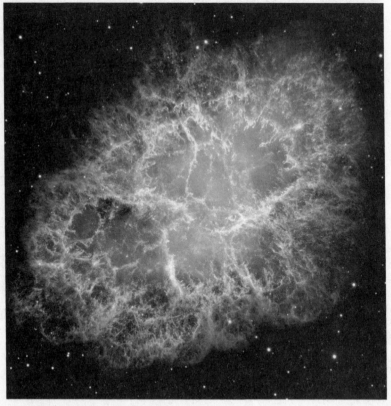

The Crab Nebula. *Courtesy of NASA*

In 2001 Japanese investigators drilled down into a glacial dome in Antarctica and recovered a cylinder of old, layered ice. Around the forty-eight-meter depth interval, they found a spike in the concentration of nitrogen oxides that matched the age of the AD 1054 star, and just below it lay another spike that matched the age of the guest star of AD 1006. Later, scientists from the University of Kansas reported similar results with ice cores from Greenland as well as Antarctica. The most likely causes of these globally synchronous pulses would be cosmic rays from supernovas that oxidized atmospheric nitrogen molecules, which in turn fell to Earth.

Here, then, was a smoking gun: The guest stars were indeed supernovas that now carry the labels SN1006 and SN1054. And you can still see some of the smoke of SN1054 in the Crab Nebula.

More clues to the origins of our sun and home planet lie back down and to the left of the Crab, in the sword hanging from Orion's belt. The middle "star" is actually many stars floating in a nebula that was built by stellar emissions and supernova explosions. The haze surrounding those stars is a broth of hydrogen and helium seasoned with powdered ice and iron-bearing mineral fragments. Recent studies suggest that most of those stars are less than a million years old, a mere blink of a cosmic eye, and some are still in the process of forming.

Stars are often born in communal nurseries such as these, which makes sense because that is where most of the raw materials for star development are concentrated. Where, then, is our own family of related stars? Our sun is strangely alone. One of its closest neighbors, Proxima Centauri, is part of a relatively small cluster about four light-years away. If the *Voyager 1* space probe, which now travels at nearly forty thousand miles per hour, were to aim for it, the trip could take up to eighty thousand years. Some experts suggest that our young sun was ejected from its own nest shortly after its birth, perhaps by the gravitational backwash of a gigantic star that cruised past it long ago.

Also within the Orion Nebula are ragged streaks in the glowing clouds of gas and dust. These are the trails of supersonic projectiles several times larger than our solar system that resemble bullets tipped with glowing caps of iron that formed in a catastrophic explosion. To a cosmologist the presence of this element is as informative as traces of gunpowder at a crime scene would be to a detective, because iron kills stars.

The source of those telltale streaks must have been much larger than the sun. It would have lived a relatively short life, on the order of several dozen million years, because supersize stars live fast and die young. And like SN1054 and the mothers of many of your own atoms, it probably died of iron poisoning. Each of these conclusions can be deduced from what we now know about the ways in which elements form in the nuclear fusion reactors of stars.

When a star is newly born it fuses hydrogen nuclei into helium, releasing huge torrents of energy in the manner that Einstein described. Our own hydrogen-fusing sun transforms several million tons of matter into heat, visible light, and other forms of energy every second. The next time you glance at the sun, try to think of it as a star that, despite its moderate size class, is large enough to swallow a million Earths. The temperature at its surface is close to ten thousand degrees Fahrenheit, and the temperature in the core is closer to 25 million degrees. No wonder it can burn your eyes and fry your exposed skin from so far away that it takes about eight minutes for the light and heat to reach you. Imagine the ferocity of the inferno at the source of all that energy, and what it is doing to the atoms in there. The blazing hydrogen that dominates the reactions dates back to the birth of the universe, but the time it has spent sizzling in the heart of the sun has been one of the most stressful periods of its history. The light and heat from the sun's fierce plasma are, in a sense, the electromagnetic cries of tormented atoms.

Cosmologists still debate the details of our presolar origins, and we may never have all the answers. But the abundance of iron on Earth anchors one piece of the tale beyond reasonable doubt. At least one core-collapse supernova seeded our stellar nursery with the cinders of its own destruction.

The source of that ancestral supernova was a star that had fused hydrogen and helium for millions of years before it began to produce new elements. Protons and neutrons clustered together into larger nuclei, and as progressively heavier clumps formed at the center of the star, concentric shells of lighter elements accumulated like the layers of an onion.

Early on in that progression, carbon nuclei formed for several centuries as temperatures in the stellar core soared to hundreds of millions of degrees. After the eventual death of the star, many of those atoms congealed into microscopic diamonds and specks of graphite, the stuff of future gems and pencils as well as the carbon frameworks of your hair, muscles, and membranes.

The fusion of helium and carbon outside the star's core produced many of the oxygen nuclei that now reside in your body water and the air that you pull into your lungs. Nitrogen nuclei also appeared in cyclic fusion reactions involving carbon and oxygen, some of which now fertilize the crops that feed you and work their way into your genes and proteins. As core temperatures reached into the billions, some of the superheated oxygen nuclei combined to produce silicon that now forms much of the rocky crust of the earth beneath you.

For several weeks silicon nuclei accumulated near the center of the star, as most of the other elements of your body churned in distinct, nested layers. Most of the nuclei of the calcium and phosphorus atoms in your bones, the sodium and chlorine in your sweat, and the potassium in your nerves were forged in just such a furnace. Meanwhile massive swirls of plasma and condensed atoms lifted away from the surface of the fireball, scattering gas and dust millions of miles out into space.

Up until that multilayered stage of development, the production of new elements released enough fusion energy to keep the star burning without collapsing under its own weight. But as the nuclei grew larger, the balance of power began to tip in favor of gravity.

Positively charged protons repel one another unless something smashes them together hard enough to make them stick. Such nuclear attraction is trillions of times stronger than gravity, but it operates only over very short distances, and the mutual repulsion of protons works more effectively against it as a nucleus becomes larger. Because of these opposing factors, the fusion of bulky iron nuclei cannot yield enough heat energy to sustain a star.

One way to envision this struggle between forces within a nucleus is to imagine pushing a heavy boulder up a steep hill, like rock-rolling Sisyphus of Greek mythology. It takes all your strength to move the boulder to the crest of the hill, but once you get there gravity can begin to drag it away from you and roll it down the far side on its own. The sudden forward motion and the release of tension from your muscles would be somewhat like the burst of energy that a nuclear fusion reaction releases in the heart of a star. But what if the final pitch of your climb becomes too slippery for your feet to grip the ground properly? You might have to run in place to hold your position, and there would be no free trip down the far side of the hill for your boulder. Worse still, if the path becomes slick enough the rock might even roll back down over you. A similarly unstable situation developed inside the star that spawned our solar system when it began to forge nickel nuclei, which quickly became iron nuclei, an energetic tipping point beyond which disastrous consequences must result.

With the reluctant protons barely holding their positions in the broad nuclei, the increasingly iron-rich plasma began to consume as much energy as it produced. Within hours to minutes of generating an Earth-size sphere of metal nuclei, the star began to smother in its own ashes until, bereft of the energy needed to resist gravity, its core suddenly collapsed.

The intense compression and temperatures crunched the tortured nuclei down into a stiff, superdense ball just a few miles wide. Atoms don't normally squash this much, but these nuclei had lost their protective electron clouds in the hot chaos of plasma, allowing them to pack so tightly that a teaspoon of them would weigh close to a billion tons on Earth. According to NASA's Chandra X-Ray Observatory Web site, the compression would be equivalent to squeezing all of humanity into the volume of a sugar cube.

The internal collapse of the star opened a perilous gap between the surface of the shrunken core and the outer envelope of plasma. Within seconds the unsupported envelope collapsed, too. It smashed against the impenetrable core and rebounded through the outer layers of the star, blasting it inside out in a stupendous explosion of heat, light, and subatomic particles. The supernova was millions of times brighter than the sun, so powerful that Yang Weide and Ibn Butlan could watch a similar outburst with their naked eyes in broad daylight from thousands of light-years away, while cosmic rays that accompanied the visible flash produced a rain of nitrogen oxides from pole to pole here on Earth.

The shock wave itself may have triggered brief fusion reactions of its own as it tore through the nebular mess of gas and dust. Some of your phosphorus nuclei could have formed in that manner, while gold, rare earth elements, and other atoms heavier than iron appeared for the first time in the short-lived blast zone. The brevity of their creation and the relative scarcity of stars large enough to produce such tremendous supernovas also makes the heavier elements uncommon—and therefore of greater value to us than more abundant substances. Because of all this, you can thank the star-killing iron in your blood for much of the inherent value of any gold items that you may possess, as well as for the relative rarity of rare earth elements.

You might also thank your blood iron for the sun. The shock wave swept up huge swarms of debris and churned it into swirling clusters with the help of energetic blasts from nearby stars. The gravity in some of the denser clumps eventually pulled in enough material to launch new fusion reactions, and some of the largest clots may have ignited as the shock front passed. As a result, the wave of death left newborn stars gleaming in its wake, with our own sun among them.

You can likewise thank iron for your home planet. Rubble from one or more iron-poisoned stars spiraled in a slowly spinning disk around the young sun as it wandered away from its birth nebula. Much of it gradually clumped together through moisture, magnetism, and gravitational or electrostatic attractions. When the early earth grew large enough, its own gravity pulled it into a spherical shape with a dense metallic core and thin, stony crust.

While you're at it, you might also thank iron for the space metal in your frying pan, your house keys, and your car, not to mention the magnetic strips on your credit cards and the compass-tugging magnetic field of the earth. And after spending more than four billion years on this planet, the petrified blood iron of your parent stars continues to play the dual roles of destroyer and creator within the blood and tissues of your own body, as well.

When Luke Combs took his pregnant wife to the University of Kentucky Hospital in 1958, some of the medical staff paid more attention to him than to his wife. Doctor Charles Behlen later remarked that the man's skin was "just as blue as Lake Louise on a cool summer day." Thus began a series of investigations that led to the discovery and treatment of a genetic blood condition that turns white people blue.

According to a fascinating and sensitively crafted report by the University of Indiana writer Cathy Trost, blue people were well known in the hill country of eastern Kentucky as far back as the early nineteenth century. When a French immigrant named Martin Fugate settled on Troublesome Creek, he fell in love with Elizabeth Smith, a woman who, by happenstance, carried a rare recessive gene that he also shared. This trait is not normally expressed in visible form until two carrier parents produce offspring, and of Martin and Elizabeth's seven children, four were born blue. So, too, were many of their descendants.

When Benjamin Stacy was born in Hazard County in 1975, his maternity nurses were shocked by the color of his skin. Perhaps fearing that he suffered from something akin to blue baby syndrome, they had him rushed to a larger hospital in Lexington for a blood transfusion. But before the procedure began, his grandmother intervened, asking the doctors, "Have you ever heard of the blue Fugates of Troublesome Creek?" Mr. Stacy recently told the Web site of Britain's *Daily Mail* that he lost his blue tint within a few weeks, although for years afterward his lips and fingernails still turned blue when he became cold or angry.

The medical basis of the Fugate blueness was worked out during the 1960s by the hematologist Madison Cawein, who also devised a safe and effective treatment for it. Cawein, who died in 1985, told Trost that he spent a summer "tromping around the hills looking for blue people" but didn't succeed in studying anyone until Patrick and Rachel Ritchie walked into his clinic.

"They were bluer'n hell," Cawein said, adding that—understandably in a color-conscious society—"they were really embarrassed about being blue." Although blue members of the Fugate lineage were just as healthy as anyone else, the stigma of their appearance and the crude stereotypes about inbreeding among mountain people could be as painful as any ailment.

Cawein's research, which was later published in the *Journal of Internal Medicine*, confirmed what earlier work on blood conditions among Alaskan Native Americans had already suggested. The blue color stemmed from iron-bearing hemoglobin molecules in the blood.

Normally bright red, hemoglobin turns muddy brown when the iron it carries loses electrons to other substances. In this form, called methemoglobin or met-H, the discolored molecule cannot carry oxygen effectively until its electron balance is restored. The blue people of Kentucky lack a functional gene for the enzyme that normally does that job, so more met-H builds up in their blood than usual.

For people who carry two parental copies of the trait in their genetic code, the condition is more extreme, and their blood color can turn from cherry to purplish chocolate. In Benjamin Stacy's case his initial color faded because he carried a normal copy of the gene, enough to reduce his enzyme deficiency as he grew. His blue tint, however, came not so much from his blood as from the tissues around it.

When viewed through pale skin, blood vessels often appear to be blue, which misleads people into thinking that deoxygenated blood is blue as well. This conversion of deeply buried reds or browns to blues also manifests itself in the so-called "Mongolian spot" that nine out of ten newborns of Asian, Native American, and East African descent display on their buttocks or lower back. The flat bluish patch resembles a bruise and is sometimes mistaken for a sign of child abuse by those unfamiliar with it (few Caucasians carry the trait), but in fact it stems from brown melanin granules that collect within the deeper layers of the skin and usually fade after infancy.

If your own skin is pale, then you can puncture the vein illusion for yourself by donating blood or by watching what happens the next time you have a blood test. When the needle pierces a blue vein, red liquid fills the sample vial. According to some scholars, this same dermal illusion may have given rise to the concept of "blue-bloodedness" among European nobility, whose creamy translucent skin was not darkened by manual labor under the sun or intermarriage with darker-skinned foreigners.

Cawein's cure for Fugate blueness was simple. All that was needed was to provide an electron donor to the blood iron, and a harmless dye did the trick. In Patrick and Rachel Ritchie's home, he injected each of them with one hundred milligrams of a biological stain called, ironically, methylene blue. "Within a few minutes," he told Trost, "the blue color was gone from their skin. For the first time in their lives, they were pink. They were delighted."

The similarity in color of the dye and the skin was purely coincidental, but it caught the attention of Cawein's patients, whom he later treated with pills rather than injections because the effects of the treatment were temporary. The dye is normally excreted in urine, and one of the mountain men reportedly told Cawein, "I can see that old blue running out of my skin."

The association of iron with blood is well known to anyone who has taken iron supplements to counteract anemia. Ocher and other rusty iron-rich minerals have also been used since ancient times to symbolize blood and life, and the name and color of hematite reflect that connection. But the actual details of how iron both serves and destroys cells are as rarely understood by most of us as they are fascinating and important to our health.

Take the cartoon character Popeye, for instance. Many of us grew up listening to his trademark refrain, "I'm strong to the finish 'cuz I eats me spinach." As parents urged generations of reluctant kids to eat their spinach too, the reasons given for it have often focused on iron through intuitive links between metal and the strength of that muscular sailor man. However, the original basis for the promotion of spinach in American culture was to add vitamin A, not iron, to the diets of children suffering from malnutrition during the early twentieth century. Somehow the message became garbled in translation.

In fact the iron content of spinach is not outstandingly high among vegetables, and the relatively modest amounts of iron found in plant foods are also less easily absorbed than the abundant iron in bloody red meat. As the hematologist Terrence Hamblin once wrote for the traditionally humorous Christmas issue of the *British Medical Journal*, "For a source of iron Popeye would have been better off chewing the cans."

Does iron really have anything to do with the strength of your body after all? Indeed it does. And in a mirror image of the strengthening of iron into steel by adding a little carbon to it, adding a little star-killing iron to your carbon compounds equips your cells to perform impressive feats of demolition, transportation, energy production, and defense.

It all begins with a single atom of iron nestled in a cozy molecular basket.

At some point in your life, by accident or design, you have surely seen what your blood looks like. Perhaps you noticed the similarity of its color to that of rust, and surmised that the iron in both substances produces the redness with the aid of oxygen. That is true up to a point, but you also owe the color of your blood to the basket-like heme molecules in hemoglobin that carry the iron for you.

A heme's five- and six-sided rings of carbon and nitrogen can resonate with and scatter light of various colors. The yellow or orange tints in a bruise come from such rings in broken pieces of hemoglobin called bilirubin, and much of the golden tint of urine stems from the further breakdown of bilirubin. Brown skin color arises from melanin molecules that likewise resemble fragments of carbonaceous netting, also without the aid of iron. Sometimes the colors of pigments represent important features of the organisms that possess them, as with melanin, which produces culturally significant skin tones while also reducing damage from the sun's ultraviolet rays. But in the case of your blood the color is secondary: The principal function of your hemoglobin is to keep you alive.

When viewed in isolation, the hemes that carry most of your blood iron resemble small sheets of chicken wire—flat, symmetrical, complex, and radiating from a distinct center. Surrounding the center are four carbon-nitrogen rings, each with a nitrogen atom pointing inward. Suspend an iron atom in that central seat and the heme is ready to shuttle oxygen from your lungs to respiring tissues anywhere in your body.

When buckled into a heme, an iron atom can reach up and snatch an oxygen molecule from solution, then release its captive again where it is needed. And when a heme is mounted on a lump of hemoglobin, the hemoglobin helps to stabilize its cargo on the rough ride through your bloodstream by using a molecular tendril to press down on the oxygen from above.

Without heme-bound iron, you could huff and puff and still not get full satisfaction from the measly amount of oxygen that your blood plasma can carry on its own. Each of your hemoglobin molecules carries four such hemes, and roughly 250 million hemoglobins pack into each of your red blood cells. With every drop of blood that squirts down an artery in your arm, several hundred million red blood cells and quadrillions of iron-bound oxygen commuters ride "the Red Line" to your hand.

Such well-behaved iron atoms have been ferrying oxygen to your cells ever since you were born, and they will continue to serve you thus until you die. In their current form, however, they would not have been able to do so as effectively while you were still inside your mother's womb. Your lungs were not working then, and there was no fresh air for you in there anyway. You had to breathe instead through your navel by extracting oxygen from your mother's blood. This required a special fetal hemoglobin (hemoglobin F) that clutches oxygen more tightly than does adult hemoglobin. That temporary trait allowed you to tap into your mother's oxygen supply through your umbilical cord until you could trade directly with the atmosphere as you are doing now.

But as important as oxygen transport is to your health, the other things that iron-bearing hemes do for you are equally essential. One potentially unsettling way in which to demonstrate this is to consider how cyanide poisoning works.

In simplest terms, cyanide kills by suffocation. The discoloration of its victims is reflected in the medical term "cyanotic," which describes the blue-tinged lips of people who run short of oxygen. In solution the lone hydrogen atom in a hydrogen cyanide molecule breaks free and turns the remaining carbon-nitrogen ion into a deadly mimic of the life-giving gas that fits neatly into the grip of a heme.

When cyanide gums up your blood iron by sticking to it as oxygen normally does, it slows the shipment of oxygen from your lungs. But even more damage occurs in some of the most remote crannies of your cells. There within your mitochondria, cyanide "SWAT teams" can sabotage the iron-bearing machinery that generates your energy.

Hemes tether reactive iron atoms to a diverse class of proteins other than hemoglobin called cytochromes ("cell color"), some of which reside in your mitochondria. Much as larger iron objects conduct electricity, those heme irons help to transmit electrons through cytochromes that process chemical energy from your food. It is this vital electron transport chain that cyanide attacks by clinging to the iron atoms, thereby making it more difficult for them to transmit electrons. Without that power source your muscles and nerves would shut down, and your lungs and heart would stop dead.

Hemoglobin is also the dominant blood protein in other mammals as well as humans, and if you have ever watched a raw steak turn brown on a grill, then you have watched hemoglobin become met-H. But the color of meat also comes from myoglobin, another iron-toting protein, which acts as a localized oxygen reservoir for muscle cells and which also turns brown when cooked. To your cells, oxygen means energy, and more heme iron means more oxygen. The strength of your muscles is therefore enhanced by the small fraction of your iron that they carry in the form of myoglobin.

Whales, dolphins, and seals rely even more on muscle-bound oxygen while holding their breath during long dives, and so do many seabirds. Their flesh is more liberally packed with myoglobin, which gives it a dark purplish hue that can be surprising when you first encounter it, as I can attest from personal experience.

Years ago I worked as an instructor at the Audubon Society's ecology camp on the coast of Maine, where I co-hosted cruises to an island where puffins had been reintroduced after overhunting and egg harvesting eradicated them a century earlier. Having become used to thinking of puffins in terms of rescuing them from extinction, I later faced an ethical dilemma on a research visit to Iceland, where puffins are harvested for food. Settling down to dinner at a restaurant in Reykjavík, I opened the menu and stared at it in horror. I was to choose between steak, salmon, and roasted medallions of puffin breast. Without going into detail here, I learned two things that evening: I can be a hypocrite when faced with temptation, and puffin meat is approximately the color of fine cabernet—which also goes nicely with it.

Roughly one-third of your iron resides in molecules other than hemoglobin and myoglobin. Iron-bearing proteins build and repair your genes, metabolize drugs and toxins, or help to construct hormones, and many of your enzymes use iron as a brutally effective cutting tool. When worn-out blood cells are due for recycling, for example, cytochromes in your liver hack them to pieces. And if you have ever applied hydrogen peroxide to a cut, you probably noticed the creamy foam that it generated. This was the work of catalase, a protective enzyme that uses four iron atoms to help it smash a million cell-damaging peroxide molecules per second into water and oxygen gas. Catalase molecules are on permanent guard duty throughout your tissues, shielding you from dangerous waste chemicals that often form inside your metabolizing cells.

A typical adult body normally contains about four grams of iron, roughly the weight of three paper clips. But just as iron can be used for benign purposes in your body, your cells can also wield it like a weapon. And sometimes an improperly handled cellular weapon can turn against its owner.

The most biologically reactive form of iron, known as ferrous iron, readily unloads electrons onto other atoms and molecules. Ferrous iron can therefore behave badly in the presence of certain compounds in your cells, forming corrosive free radical molecules that damage tissues or disrupting the clotting of blood in a wound. Although iron is helpful to you when it operates in a controlled fashion, a single gram of rogue ferrous iron is enough to hospitalize a small child. A typical lethal dose, most often the result of children being fed adult doses of iron supplements, can range between three and six grams (the average adult lethal dose is somewhere between ten and fifty grams). According to a report in *Pediatrics*, sixteen American children below the age of six died in this manner between 1983 and 1990.

Why hasn't the iron in your blood killed you already? You normally keep most of it tied up in hemes and other molecular tethers, and your cells employ an army of handlers to make it do your bidding with minimal collateral damage. The most common of these molecules are tangled proteins that can hunch over an iron atom or engulf it. Chief among them is ferritin, which can sequester iron inside a cell, and transferrin, which transfers iron between cells.

The destructive powers of your iron atoms can, of course, also be turned against pathogenic microbes. One of your first lines of immune defense is a kind of scorched-earth policy that deprives invaders of food and supplies. At this very moment you are fighting bacteria simply by using hemoglobin and other substances to lock up the iron in your blood, which denies them access to the atoms they need for their own enzymes. This not only protects you from the dangerous side effects of free-floating iron but also helps to prevent pathogens from using it against you. Your red blood cells normally confine your hemoglobin, but when they eventually break or wear out, any iron leakage is quickly soaked up by ferritin and transferrin. This leaves your body virtually devoid of reactive free iron, which is generally a good thing because you don't want much of it floating around inside you anyway.

That unusual scarcity of free iron inside you also has a downside, however. It can betray you to microbes. After sensing suspiciously low iron levels in contrast to richer conditions in the environments outside your body, formerly dormant bacterial genes suddenly switch on. This unleashes swarms of proteins that can steal your iron and give it to an invading army.

Some of these proteins puncture your cells in order to release their iron-bearing molecules. Other specialized proteins on the surfaces of the bacteria tear the exposed iron atoms away from their guardians or, in some cases, consume the molecules whole. Yet another class of bacterial products, called siderophores ("iron bearers"), are among the strongest iron-binding substances known. They act like edible sponges that absorb and hold iron compounds until the hungry microbes can engulf them, iron and all.

The presumed links between supernovas and plagues suggested by the Persian scholar Ibn Butlan are, in fact, borne out on this atomic stage. *Yersinia pestis*, the bacterium responsible for bubonic plague, deploys a siderophore called yersiniabactin, which it uses to steal iron from its victims. So effective are siderophores that nonpathogenic bacteria and even many plants use them to coax iron away from rust particles in soil.

In response to this attack, however, your own cells can launch a second wave of defense. Lactoferrin snatches up free-floating iron debris and kills bacteria by boring holes into them. Siderocalin leaps onto the siderophore iron sponges, covering them up so the bacteria can't find or absorb them. The conflict escalates further as the invaders release more devious kinds of siderophores that can't be clung to as easily, and it all becomes very complicated as inflammatory responses and fevers turn your body into a battleground. Some bacteria, such as the *Borrelia burgdorferi* that cause Lyme disease, simply bypass the confusing complexity of these iron wars altogether by using less aggressively targeted or protected manganese atoms for their metabolism instead.

Clearly, then, iron is as important to you as it is potentially dangerous. The same is true for other living things for similar reasons, which accounts for the presence of biological iron in your food. Mammal and bird flesh are full of hemoglobin and other iron carriers just as yours is. Although they supply you with less iron than a Popeye cartoon might seem to suggest, spinach leaves and other veggies also contain that element because they use it in their own cytochromes and various free-floating enzymes. Planktonic algae can mine iron from bacterial siderophores at sea, and bacteria that live in the roots of alfalfa and other legumes use iron-laden enzymes to produce fertilizer from nitrogen in the air, while also jointly creating with their host plants a version of hemoglobin that helps the subterranean cells to breathe.

Your elemental iron links you through your food to the soil from which plants draw their iron, and then on back through ancient rocks of the earth's crust and asteroid dust to the explosive birth of the solar system itself.

Throughout history people have benefited from iron's properties, not only in blood and muscles but also in Earth's magnetic field, which shields us from cosmic radiation and prevents the solar wind—the sun's fierce blasts of energy and subatomic particles that a student of mine once likened to "a big angry hair dryer"—from blowing our atmosphere off into space. With the advent of iron smelting and processing technologies, however, we have more recently added artificial extensions of star-born metal to our bodies just as early animal ancestors developed shells and skeletons from other elements.

Originally only scarce meteoric iron was used for adornment and tools until serious Iron Age smelting began in Asia and Africa. The new technology was widely used in Anatolia and possibly also in tropical Africa three to four millennia ago, reached Europe and China between 3,000 and 2,500 years ago, and spread throughout the rest of the world through trade and warfare. The Iron Age was already old news by the time of Yang Weide and Ibn Butlan, and by most accounts it has long since been replaced by the industrial age, the space age, and the digital age. But iron is still by far the most widely used metal on Earth, mostly in the form of steel. According to the World Steel Association, China alone produced more than seven hundred million metric tons of steel in 2012, about half the global total and quite a change from the few hundred thousand tons of iron per year that were processed around the time of guest star SN1054. Magnetic iron is now used in satellites and GPS units, recording devices and stereos, the pickups of electric guitars, and the hard drives of computers. It is becoming increasingly difficult to imagine modern civilization without electromagnetic media, but even before they came along we were already more dependent on stellar iron and its alloys than we were a thousand years ago.

Notice the various snaps and zippers on your clothing, which was sewn with sturdy steel needles. Notice, too, the scissors and staples in desk drawers; the hinges, doorknobs, and locks that let you in and out of rooms; the nails and screws that hold houses together; and the steel skeletons that hold skyscrapers upright. Your food is prepared and consumed with the aid of iron-based utensils, transported over steel-supported bridges on steel rails and truck beds or in steel-hulled ships, and grown in steel-plowed fields. Police and military personnel carry steel weaponry in your defense, using modified versions of the cast-iron "fire-tubes," "bandit-striking penetrating guns," and "flying cloud thunderclap eruptors" of ancient China. And industrial steel tools cut, pound, pull, pour, drill, and stamp most of the objects on the foregoing list, while the heat-trapping carbon emissions from the coal required to process all that iron ore changes the chemistry and temperature of the atmosphere.

As iron influences stars and cells in both positive and negative ways, it will also continue to serve or harm us through our ever-more-sophisticated implements of creation and destruction. How we strike that balance will depend upon our ability to understand our atomic connections to one another, to this planet, and to the deep black sea of space that we all cruise together.

Our machines now carry some of us into orbit and extend our senses beyond the solar system, and formerly inaccessible realms are coming to seem more like part of our home territory. We can now better recognize the flash of a "falling star" for what it is, a piece of our cosmic neighborhood that doesn't actually fall but merely collides with our atmosphere and then mixes its atoms with air, water, soils, and perhaps even our blood. And when we step outside at night to view the moon and stars we can now better understand that the stuff of distant worlds is also here on Earth with us and within us. As we continue to build upon the work of our scientific forebears, how we interpret and respond to such things will help to chart our own unique course through history.

4

Carbon Chains

*Our atom of carbon enters the leaf . . . (where) like an
insect caught by a spider, it is separated from its oxygen,
combined with hydrogen . . . and finally inserted in a
chain, whether long or short does not matter, but it is
    the chain of life.*   —*Primo Levi*

*Organic chemistry is the chemistry of carbon compounds.
Biochemistry is the study of carbon compounds
    that crawl.*   —*Attributed to Mike Adams*

How would you like to be made of air pollution? Believe it or not,
much of your body already is.

If you could imagine nine out of ten of your atoms suddenly losing their color, the remaining tenth would leave a ghostly translucent
form of you behind, as though you were sculpted from smoked glass.
That's what you would see if you could envision the eight hundred
trillion trillion atoms of carbon that are now embedded in your body.
Incredibly, about one in eight of those carbon atoms emerged recently from a smokestack or an exhaust pipe.

By the end of this century, rising carbon dioxide emissions from the
burning of coal, oil, and natural gas will lodge even more fossil carbon
in the bodies of your descendants. This is because plants harvest carbon dioxide from the air, and when we dump fossil fuel carbon into the
atmosphere much of it runs through the world's food chains and into
our bodies. If we continue to burn fossil fuels at current rates until the
remaining exploitable deposits run out, then people of the next

century and beyond will be "children of the fumes," a species whose organic matter is largely derived from its own waste gases.

Such claims may seem outlandish, but they are not. Although we are atomic beings whose individual particles have resided in other people and places that we may enjoy thinking about, the same connections also tie us to less pleasing things, as well.

As I write these words, carbon dioxide comprises an average of 400 out of every million molecules in the air around me. When my ancestors first came to America from Germany during the eighteenth century, they breathed air that carried less carbon dioxide, closer to 280 molecules per million. The extra dose that we now inhale is traceable to the combustion of fossil fuels during the last two centuries or so. These amounts are minuscule compared with oxygen, which is thousands of times more abundant, but that puny fraction of oxidized carbon can melt vast ice caps, derange the chemical balance of the oceans, and also become an ever greater part of you.

When I called the atmospheric chemist Ralph Keeling at his laboratory in La Jolla to ask about the fraction of carbon in our bodies that comes from fossil fuels, he pointed out that although one-quarter of today's airborne carbon dioxide is closely linked to human activity, some of the linkages are more direct than others. "Gases are always moving into and out of the surfaces of the oceans," he explained, "so carbon dioxide molecules that were already present in seawater can easily trade places with some of the ones we add to the atmosphere." In his estimation the rise from 280 to 400 ppm CO_2 represents a roughly equal blend of our emissions and former marine molecules that fossil fumes have displaced from the oceans. This means that about half of the additional carbon in modern food chains—or about one-eighth of the carbon in your body—came from the carbon dioxide exhaust of fossil fuels, and the other half represents refugees displaced from the sea.

In ages past, the connections between food and wastes on small farms were more obvious to the average person. Local fields produced hay for cows, which fed people and fertilized those same fields with manure. The rise of industrial civilization and the growth of cities have made it more difficult for people to see where their food comes from and where their wastes go. But today new scientific findings are helping to close that cognitive gap again by revealing the atomic nature of existence. As we learn to follow the comings and goings of elements within and around us, we can begin to ask some profound questions about our connections to the world. What does it mean to live in a body that is partly made of carbon waste? Where exactly do your own personal carbon atoms come from, what do they do inside you, and where do they go? Such questions can lead to surprising answers.

In simplest terms you build most of your body from air and water, but to do this you need the help of plants and light-harvesting microbes. Every protein fiber in your muscles, every fleck of body fat or speck of blood sugar, every gene, membrane, and bone in your body bears a framework of carbon atoms that were pulled from the same atmosphere that surrounds you now, in which carbon is as rare as gold dust in a gravelly streambed. But that precious fraction, when sifted out and melded into plant sugars, can also be reworked into other living things, including you.

Carbon invades your lungs with every breath, but it only becomes an integral part of you when you eat or drink it. You, in that sense, are a body snatcher, a stealer of carbon from other living things. If you trace your atomic supply lines backward from fork to farm and fishnet, you will eventually meet plants, algae, and cyanobacteria. These are your portals to the airborne carbon reservoir—the primary producers of the planet.

Although it is easy to say that "all life is connected" in some vague sense, most of the connections are hidden from casual view by size, time, and distance. Their invisibility makes it easy to overlook them and more difficult to understand or believe in them. Even if you grow or hunt your own food, it isn't normal to think of yourself and your surroundings in atomic terms. So here is an invitation to be abnormal for a moment. Consider, if you will, your inner carbon.

A single-file line of twenty million of your carbon atoms could barely encircle a poppy seed, and trying to see any particular one of them would be akin to staring up at the moon and trying to spot an astronaut's bootprint on its powdery surface. However, if enough carbon atoms gather together in one place, they can produce effects and objects that are much easier to perceive.

You can feel carbon compounds bumping into you whenever a breeze brushes your cheek, for example, and there are a lot more where those came from. Roughly a million trillion trillion trillion CO_2 molecules wander the atmosphere, each one of them a carbon atom suspended between two bulbous wings of oxygen. A paper in *Science* by the oceanographer Paul Falkowski and his colleagues reported that the oceans carry about fifty times more carbon dioxide than the atmosphere, and two to four times as much carbon occupies the living tissues of organisms as rides the air. At any given moment the average adult has thirty to forty pounds of them embedded in his or her body.

A purified lump of carbon can look like black crud, but strangely enough the same dark dust that you might wipe from the palm of your hand after cleaning a chimney or cookpot also resides within your hand. The carbon inside your hand, of course, doesn't look like crud at all while you're still alive, but it would if you extracted it from the tissues that it comprises.

This is an important point. The carbon atoms in your hand are identical to those in soot, and some of them may actually become soot someday if your mortal remains are cremated. They are also identical to the carbon atoms in everyone you have ever met or heard of, as well as in every imaginable creature from worms to wombats, the plastic of your computer screen, the gasoline that powers your car, and the asphalt that you drive on. Carbon atoms are much like LEGO blocks in that they can be stacked and linked in endless ways to form both the simplest shapes and the most intricate flights of fancy. The key to their versatility is in the nature of the connections that they make with one another and with other elements.

A typical carbon atom carries within its nucleus a dozen protons and neutrons, and it also has four electron-cloud connectors that can harness it to other atoms with strong covalent bonds. Any single carbon atom can double-grip two atomic partners or join a long molecular chain while carrying two other atoms with it. In this manner the same carbon atom can become part of an almost limitless number of molecules, moving from role to role like an actor auditioning for different parts in a theater. The carbon atoms in your eyelashes could just as well help to produce the transparent corneal proteins and distinctive pigments of your eyes. In ages past they may also have been spun into the world's first spiderwebs, or brightened the colorful wings of ancestral birds, or sweetened the fragrances of prehistoric flowers.

Recognizing the ephemeral nature of carbon compounds helps to make the fossil fuel atoms in your body less troubling than they would be if they retained their formerly smelly, smoky properties inside you. Look far enough back into history, and you'll find your former coal and oil carbons in lovely forests and pristine oceans of the distant past. Atoms abandon the traits of their earlier forms when they take on new ones, so it makes little or no practical difference where the carbon atoms in your breakfast cereal recently resided. For them every new molecular arrangement is a new beginning.

The atoms in your body are not distracted by memories of earlier times spent in cave-bear breath or other previous incarnations, and the weaving of waste carbon into your flesh has no direct effect on your health or appearance. Human bodies have always been cobbled together from recycled pieces of people and other things that have used and discarded them over the ages. We are all, in that sense, the "living dead." But our recognition of such atomic connections can make it easier to remember what many of our ancestors knew in more intuitive ways: None of us lives in true isolation from the rest of the world. As seventeenth-century poet John Donne wrote, "No man is an island, entire of itself; every man is a piece of the continent, a part of the main."

In earlier times people could more easily sense their connections to land, water, and air, and could thereby more readily identify the physical constraints on their lives. Today we seem to forget that supermarkets don't create the goods they contain, and we act as though our drains and trash receptacles were bottomless black holes.

In such a setting recognizing the presence of fossil fuel atoms in your body is a valuable exercise, not because they hurt you directly but because they link you to climates, habitats, and people all over the world. More worrisome, though, are the harmful substances that may accompany those carbons and become part of you, too.

Coal often contains small amounts of mercury, for example, and the U.S. Environmental Protection Agency (EPA) estimates that coal-burning power plants in the United States alone released more than fifty tons of mercury into the atmosphere each year since 1990. Toxic mercury is now so widely dispersed that swordfish from the open ocean typically carry a mean concentration three times higher than what the EPA considers to be unsafe for human consumption (0.3 ppm). Another EPA study recently found that more than a million American women of childbearing age carry unhealthy concentrations of methyl mercury in their blood serum and hair, thereby putting more than 75,000 newborns at risk of neurological damage from the mercury they picked up in the womb.

The skewed carbon balance of our bodies, then, shows that we inhabit a finite world within which we are all deeply embedded. In this realm of atoms, material things are not truly created or destroyed but are instead recycled and reconfigured. And when it comes to dealing with our wastes while we go about the business of living, it is good to remember that, at least here on Earth, there is no such place as "away" for us to throw them into forever.

Despite the astonishing diversity of carbon compounds, one feature that they all have in common is the ability to be unmade. This can happen through microbial decay or digestion, but the simplest approach is through heating. Toss any carbon compound into a hot-enough furnace and it will oxidize into wisps of gas. Even a diamond will burn if you can afford to sacrifice it for science, and the British chemist-physicist Michael Faraday described just such an experiment in the early nineteenth century. "The diamond glowed brilliantly with a scarlet light," he wrote, "inclining to purple and, when placed in the dark, continued to burn for about four minutes."

Rather than ask you to burn a diamond, I recommend setting fire to some inexpensive organic substance in order to watch this atomic transformation unfold, just to demonstrate the principle. Sure, you have seen it happen before in a fireplace or at the glowing tip of a cigarette, but observing it mindfully is another matter. I often do this sort of thing with my students in order to demonstrate the elemental nature of life. One of my favorite ways to do it is to step outside and walk over to the nearest balsam fir tree, of which we have many here on the beautiful wooded campus of Paul Smith's, the small college in upstate New York where I have taught natural sciences since 1987.

The last time I demonstrated this point, I snapped a dry twig from the end of a branch and said, "See this piece of tree? It may not look like a pile of carbon atoms, but it is. Let's unveil them with the magic of . . . does anyone have a lighter?"

The twig crackled and smoked while flakes of charcoal and ash fell to the ground, and I watched the faces of my students for signs that they recognized the underlying lesson of the demonstration. These same atoms are also what we are made of, and in time we will all share the same fate as the twig, if not through flames then by slower processes of decay.

"We've just unraveled a season's worth of sunlight, raindrops, and mountain air," I continued. "This dark char is purified carbon with most of the other atoms driven off. The ash is the minerals that the tree soaked up from the soil. But most of the twig is now blowing off into the woods as carbon dioxide and water vapor. Those trees downwind of us are soaking up some of the carbon now and turning it into new twigs. When deer pass through here later on, maybe one of them will snack on those twigs, and next year one of you hunters might turn it into a meal that becomes part of you for a while."

You never know, of course, how students will respond to such a moment. As I paused to let the information sink in, one young man spoke up.

"If the trees can turn carbon dioxide into twigs, are they turning your breath into twigs, too?"

"Sure," I replied. "Excellent! You're getting the point, I see."

"So if I eat my deer next fall," he continued with a mischievous grin, "does that count as studying, because I'll be absorbing some of the hot air from this lecture?"

Needless to say, we often enjoy an informal atmosphere here at Paul Smith's. But as we laughed at the joke, I knew that the humor was this guy's way of showing that he had glimpsed his own atomic nature. His latter surmise was, after all, correct.

Herein lies a revolving door between living and nonliving matter. On land, plants pull carbon dioxide from the air and knit its atoms into sugars that become cellular building blocks and energy sources. Some specialists estimate that about 120 billion metric tons of carbon—roughly equivalent to one-sixth of the atmosphere's entire gaseous carbon supply—cycles through leaves and stems in this manner every year. Much of it diffuses back out into the air untouched. But the carbon that becomes sap or seeds is likely to return to the air before long as well, either as waste CO_2 from the plants themselves or from things that eat plants.

To put it more simply, the atmosphere becomes us and vice versa. Our cells weave its invisible atoms into our bodies and then release them again into a fantastically rich, interlocking economy of life in which the common currency is carbon.

Most of the carbon compounds that you eat become fuel for motion and body heat, with only about one-tenth of them becoming building materials. In the metabolic factories of your cells, carbon atoms break loose from food molecules and drift free as carbon dioxide to be discarded with each breath. But your intake of oxygen is not so directly tied to that release of carbon fumes as it would be in a furnace.

The cycle of your breathing is a more complex duet between the consumption of oxygen by one set of processes and the dismantling of oxygen-bearing food molecules by another. As a result most of the oxygen atoms that escort carbon out of your bloodstream are not the ones that you just inhaled, as you might surmise from watching a fireplace draw air from a room while sending smoke up a chimney. Your incoming oxygen atoms become metabolic water instead, and most of the oxygen atoms in the CO_2 of your breath entered your body through your stomach, not your lungs. Because of this, you are more likely to exhale the atoms in the last slice of bread you ate than to excrete them, as your cells unravel them back into the breezes from which they were gathered in some sun-drenched wheat field.

You also exhale yourself, as well. The cells of your body undergo continual repairs and replacements, so many of your carbon compounds end up in the jaws of your own digestive enzymes and emerge from your lungs amid the waste gases of processed food.

From an atomic perspective you are a wondrously complex form of condensed air. It should be no surprise, then, that changing the composition of the atmosphere can alter the composition of your body, and in today's increasingly crowded, industrialized world we are doing just that. We are not only the producers of air pollution: To an ever-increasing extent, we *are* air pollution.

Diane Pataki knows just how sensitively air can respond to the waste gases released by large numbers of people. As an environmental biologist at universities in California and Utah, she has helped to launch a new field of ecological research by tracking the movement of atoms within a new kind of ecosystem—the megacity.

Among Pataki's many publications is one that appeared in *Geophysical Research Letters* in 2006, which described how the air in and around a large city responds so quickly to the behavior of its residents that its atomic composition changes from hour to hour depending upon what the people are up to. In documenting our effects on local air in this manner, Pataki and her colleagues also help to show how we affect the atmosphere as a whole.

The ecosystem under scrutiny was Salt Lake City, Utah, but other studies have found similar processes at work in other urban centers, as well, including Los Angeles, Phoenix, Baltimore, and Paris. One of those processes, which until recently drew little attention from ecologists whose work traditionally focused on wilder places, is the creation of a bubble of carbon-rich gases by city dwellers and their machines—what Pataki and others call an "urban CO_2 dome."

The term is a bit of a misnomer. An urban CO_2 dome has no firm shape and few sharp boundaries between inside and outside. It is rather like a balloon full of carbon-enriched breath minus the rubbery skin, and because its contents are generally heavier than air they tend to hunker down over the city in a great amorphous lump. The fuzzy borders melt off and dissipate downwind as the contents of the blob are continually recharged from below.

Every megacity has one of these domes, but the size, shape, and composition of each one are as unique as the environment and culture that produce it. The CO_2 dome over Los Angeles, for example, is molded by westerly Pacific winds that press it into the lap of the surrounding mountains, and it is capped by a lid of overlying air that keeps the urban gases from escaping upward. It inflates and deflates rhythmically, swelling to half a mile or so in altitude when the sun heats the air within the bubble, and collapsing to about half that height at night.

In the more open setting of Salt Lake City the dome is less rigidly confined, which allows it to wobble and stretch more freely in response to the weather. But it is what happens inside the dome that most clearly reveals the atomic connections between people and air.

Between December 2004 and January 2005 Pataki and her colleagues analyzed the chemistry and isotopic balance of the air in and around Salt Lake City and found that sudden bursts of CO_2 typically flooded the dome twice a day, first in the predawn hours and then again during the late afternoon and early evening. By analyzing distinctive isotopic fingerprints in the fumes, the team was also able to trace the extra carbon atoms in the dome to the increased combustion of fossil fuels at predictable times that reflected the lifestyles of local residents.

Most of the first pulse came from natural gas heating units during the chill of night, and the second arose when commuters choked the roads during rush hour. Surprisingly, however, as much as half of the carbon dioxide sometimes came from sources other than people and their contraptions. On warm summer nights the trees, lawns, gardens, and even the partially exposed soils of a major city can also release large amounts of CO_2 through microbial decay and plant respiration. And when the sun comes up again and photosynthesis by the local vegetation kicks in, CO_2 levels can drop within an urban dome much as they do in the planetary atmosphere as a whole in response to the cycle of the seasons.

A Web site titled $SLCO_2$ posts data from the University of Utah that show that CO_2 concentrations over Salt Lake City are usually highest on weekdays, probably because there are more cars on the road then. Carbon dioxide levels rise during daylight hours when people and machinery are most active, and weeklong readings from various monitoring stations resemble a choppy ocean. The highest concentrations develop when people turn their furnaces on and winter's cold keeps the dome from inflating fully, which traps enhanced emissions inside a smaller bubble.

As you might expect, carbon dioxide is especially abundant where the most people congregate. Scientists from Boston University recently found that CO_2 concentrations in their home city generally ranged 20 or more ppm higher than in rural Harvard Forest (averaging about 393 ppm), an hour and a half drive to the west. A study of the air in the bustling center of Cotonou, Benin, in central Africa measured CO_2 concentrations more than twice as high as the world average thanks to traffic, household emissions, and industry. And scientists who monitored the air over Indianapolis were recently surprised by a sudden increase of local CO_2 concentrations until they remembered that the Indianapolis 500 auto race had just been held. It wasn't the race cars that caused the spike but the vehicles driven by the hundreds of thousands of spectators who came to town to watch the Indy.

How does all this relate to life under such domes, where more than three-quarters of all Americans now reside? Toxic pollutants in big-city air are well-known threats to human health, but they are not the focus of attention here. This is a story about the carbon atoms in those great bubbles of air, not the potentially nasty compounds that some of them produce. And the most common of those compounds, colorless and odorless carbon dioxide, isn't abundant enough in an urban dome to harm anyone physiologically.

In fact the double load of carbon in Cotonou's CO_2 dome actually seems to be making local plants thrive. According to the aforementioned study, grasses within the most polluted parts of the city are using exhaust fumes as food. Sniff a flower in Cotonou, and a fifth of the carbon atoms in those fragrant molecules will have emerged recently from nearby cars, trucks, and chimneys.

Although plants may revel in the wealth of airborne carbon atoms in the CO_2 domes of our megacities, we don't. When we inhale carbon dioxide, we merely blow it out again along with an additional measure of our own. But that is not to say that urban carbon doesn't affect us at all. Far from it.

Most of the carbon in the CO_2 dome of a megacity comes from outside air, the same stuff that you could find on a South Pacific island or the Arctic tundra. Most of the carbon dioxide in the atmosphere, in turn, comes from respiration in the biosphere along with some inputs from volcanic vents and the ocean surface. But about one-quarter of it is traceable to fossil fuel emissions, most of which emerge from urban CO_2 domes.

This is one of the reasons why the work of people like Diane Pataki is so important. When it comes to heat-trapping carbon emissions, urban areas are major conduits between us and the atmosphere. Carbon dioxide concentrations have been rising worldwide since the Industrial Revolution, and by the end of this century they are likely to have doubled relative to pre-industrial times. Although some still deny that people are behind this rising tide of carbon, the numbers show otherwise.

Yes, fires and the decay that accompanies deforestation release CO_2 into the air, but that only accounts for a minor portion of the increase during the last century. Furthermore, isotopic analyses show that the main sources of the increase are fossil rather than living carbon.

Yes, there are many more of us respiring into the atmosphere today than in the past, but human breath represents only a tiny fraction of the carbon emissions within an urban CO_2 dome, typically on the order of 1 percent, and it amounts to virtually nothing on a planetary scale.

Yes, volcanoes spew carbon dioxide, but even gas-gushers such as Pinatubo in the Philippines or Mount Saint Helens in Washington State are localized part-timers. We, on the other hand, are at it globally and relentlessly. In an article that appeared in the geoscience newsletter, *EOS*, the volcanologist Terrance Gerlach described the more than hundredfold difference between volcanic and human emissions in stark terms. According to his calculations, we release in three days what all of the world's volcanoes typically release in a year. Gerlach's numbers also show that seven hundred Pinatubo-style eruptions would barely equal our current annual carbon dioxide output of roughly thirty-five billion metric tons, and the 1980 Mount Saint Helens eruption would have to occur at least nine times per day throughout the year to match us.

As Gerlach put it, measuring ourselves in this manner against some of Earth's most powerful geological forces "is a telling perspective on the size of humanity's carbon footprint." And to put it another way, the ceaseless flow of carbon into the air from coal, natural gas, and petroleum is beginning to turn the entire atmosphere into one enormous CO_2 dome.

The sheer magnitude of our carbon emissions is impressive, but mostly in an abstract, intellectual sense. More striking is to see exactly where most of the action is on a map.

A National Oceanic and Atmospheric Administration Web site called CarbonTracker CT2011_oi displays colorful maps based on data from satellites that monitor carbon dioxide concentrations all over the world. Click one link, and you can watch the American heartland turn blue in summer as crops and forests pull carbon from the air, then go red in winter as microbial decay in the soil releases CO_2. Click another link, and you can watch vegetation and plankton turn the hemispheres alternately blue and red with the passing seasons. The images are so detailed that you can watch regional blobs of carbon dioxide swirl and mix into the atmosphere like cream in a cup of coffee.

To watch what the intrigues of human societies can do to the air above them, check out the U.S. Energy Information Administration's graph of carbon dioxide emissions in the United States during recent decades. They increased from year to year until about 2009, then plummeted while the American economy went into a tailspin. As millions of cash-strapped people cut back on travel and consumption, their carbon emissions shrank, too. Economists and politicians might do well to consult such charts when judging the health of a nation, as a physician monitors a patient's respiration. The reduced carbon outputs during the Great Recession were analogous to the shallow breathing of a sick economy. At the time of this writing, however, the American CO_2 curve is beginning to creep back up to its former levels as the economic system slowly recovers.

From perspectives such as these, the local blends seamlessly into the global. As Keeling and others have shown, the lower atmosphere mixes completely within a year or so, and monitoring stations worldwide measure similarly rising concentrations of well-stirred CO_2. Life on Earth continues to ripple that rising line with seasonal pulses of mass photosynthesis and respiration as alternating halves of the biosphere inhale during summer and exhale in winter. But the global pool of carbon dioxide is also swelling overall, thanks for the most part to a few regions of the Northern Hemisphere.

A map posted by the Carbon Dioxide Information Analysis Center identifies those hot spots in vivid color. When seen from above, as it were, the eastern half of the United States, northern Europe, and much of India and China resemble smoldering fire pits. Within those hot spots are the world's largest urban-industrial CO_2 domes, the primary spigots that are flooding the atmospheric pool with fossil carbon. Because these sources cluster within the northern mid-latitudes, air-sampling stations now register slightly higher CO_2 concentrations in the Northern Hemisphere than they do in the south. Before the age of fossil fuels, upwelling currents in the southern oceans enriched the southern atmosphere relative to that of the northern half of the planet, which has more land and less ocean on it. Today the carbon imbalance is reversed, simply because the north contains more cities.

You probably already know that carbon dioxide warms the atmosphere. Perhaps you have heard how the arrangement of carbon and oxygen atoms tunes a CO_2 molecule to the vibrational frequency of infrared radiation and gives it powerful heat-capturing abilities. Information about the subject is easy to find, so if you need a refresher on what the greenhouse effect is and how the buildup of CO_2 is changing your world there are plenty of resources to be had.

But these pages are more concerned with the atomic connections behind the story of global warming than its ill effects, which can be depressing enough to make you want to jump from the top of an urban CO_2 dome. Following those connections leads to a deeper appreciation of what we are doing to our home planet—and, by extension, to ourselves and our descendants—but it also reveals some uplifting things as well. In the case of carbon, the atomic perspective helps to show how intimately wedded to the earth we are, for better or for worse.

When we talk about fossil fuels such as coal, oil, and gas, we really are talking about fossils.

If you heft a lump of coal, for example, you hold in your hand the carbon atoms of long-dead trees, herbs, and mosses, a dense bouillon cube of organic matter distilled from a primeval bog or swamp-forest. That's why coal burns, after all: On the atomic scale it is like refined firewood. Over geologic time, slow decomposition and the heat and pressure of deep burial reconfigure the plant remains into interlocking rings of carbon atoms that outnumber the other elements of life by as much as nine to one. A dense molecular honeycomb forms, and what was once brown or green fibrous tissue becomes increasingly dense black stone. Sometimes, the ghostly flattened forms of the original plants remain intact with such detail that specialists can identify their species.

Oil, in contrast, is a stew of carbon-rich strings that slither about in liquid form. In the case of petroleum (the word translates to "stone-oil"), the primary carbon source is marine plankton. Rather than store photosynthetic sugars and starches that can dissolve or swell up when moistened, many algae stash their carbon in droplets of water-repellent oil, the microbial equivalent of body fat. When they die, their cells sink to the seafloor, where they may be squeezed, cooked, and purified into flammable liquids.

Natural gas molecules, the last of the three main groups of fossil fuels, are the most mobile of all. A methane molecule, for example, is a single carbon atom with four hydrogen atoms stuck to it, and propane is a short strand of carbon that is likewise studded with hydrogen. Natural gas is released by bacteria that live amid the recently buried botanical or planktonic dead until increasing heat and pressure kills them, too, and it can also leak from coal seams and petroleum pockets.

The upshot of all of this is that fossil fuels are, on the atomic scale, just as much a product of the biosphere as we are. The main difference is that their carbon atoms have been separated from ours by millions of years of history and thick layers of rock. Fossil fuel carbons are like long-lost relatives who wandered off one day and were never heard from again—until now.

At least twice as many fossil carbons are thought to lie buried as there are in active circulation today, representing a huge organic bank account that built up over the ages in geological deposits. Imagine if they all came back to life, like Lazarus from the grave. Well, you don't need to limit yourself to imagining this situation. You can actually see, feel, hear, taste, and smell it, because it is happening all around you and even inside you.

Many people share the opinion of Norman Mailer, who once wrote about petroleum-based products: "We divorced ourselves from the materials of the earth, the rock, the wood, the iron ore; we looked to new materials which were cooked in vats, long complex derivatives of urine which we called plastic. They had no odor of the living . . . their touch was alien to nature." But a more nuanced look at the modern world from an atomic perspective shows that the elements of land, sea, and air still permeate our lives, and even what appear to be the most artificial, unnatural aspects of urban existence are more closely linked to wild nature than you might imagine.

To illustrate this, let's visit McDonald's.

Love them or loathe them, those Golden Arches are everywhere, and with billions and billions served, the McDonald's brand has become an icon of Western culture since its start in a lone barbecue restaurant in San Bernardino, California, during the 1940s. It is also a prime habitat for carbon watching.

In this thought experiment, you drive into the parking lot and switch off the ignition. Your car's engine stops burning the strands of algal carbon that we call gasoline, and you step out onto the hard black surface. The asphalt is a slurry of carbon strings and rings that sun-soaked marine plankton wove millions of years ago, with some pebbles mixed in to stabilize it.

As you walk up to the counter to place your order, the petroleum-based floor tiles shine with planktonic wax. Polymers of algal carbon in your shoes squeak softly as your feet press them against the slick surface. On the wall before you colorful panels display the menu through transparent panes of carbon compounds whose atoms were harvested from the air and oceans by primordial plankton. Electricity that was generated in power plants fed by Paleozoic forest carbon keeps everything brightly lit.

"A small order of fries and a Coke, please."

Sticks of botanical carbs gurgle in carbon-based oils that were made by corn, peanut, or soybean plants, transported here with vehicles that run on refined algae (gasoline), and then cooked with the help of energy from Appalachian paleo-forests (coal). They come to you in a crinkly bag of pulped tree fibers (paper), ready to be doused in the crushed carbon-rich fruits of tomato plants (ketchup). The fizz in your soda was produced by reacting microbial exhalations (natural gas) with water vapor, and the soft drink's sweetness comes from the sap of tropical grasses (sugarcane).

You pay the cashier with—what else?—a thin, flexible slab of algal carbon that was unearthed from the Arabian desert, and you're on your way.

A modern world isolated from nature? Hardly. What appears to be isolation is an illusion that a firm grasp of atomic reality quickly dispels. Even here, in what is arguably one of the most human-centered, artificial places on Earth, carbon atoms connect you to a wealth of organisms from today and the distant past. In this sense plastics and their chemical kin are no more unnatural than wool (processed grass molecules by way of sheep) or wood (processed air and water molecules by way of trees).

From a strictly atomic perspective, a bustling city center can have a lot in common with, say, a remote tropical island. The windows of the vehicles and buildings are sheets of molten beach sand, and the walls, foundations, and sidewalks are crushed limy minerals drawn from seafloors of the distant past. Let your imagination run with such facts, and you might even envision the skyscrapers as reincarnated versions of coral reefs complete with filter-feeding storefront "polyps" that draw their sustenance from the plankton of people streaming past them. All of these structures and artifacts consist of recycled materials that we rework to our own ends just as microbes, plants, and animals have done for millions of years.

This is not a traditional way of thinking, of course, and it might easily be misunderstood, so a brief clarification may be appropriate here. Using atoms to expose the illusion of isolation from nature in this manner is not a license to turn the whole world into a paved playground for humanity; far from it. If anything, it serves to remind us that as atomic beings we are no more exempt from the laws and processes of nature than any other species, and that we ignore them at our own peril.

The elemental connections between you and your food run ever deeper into the past as more and more fossil carbon returns to general circulation. Take that ketchup, for instance. If the tomatoes were grown in the Central Valley of California, then many of their carbon atoms came from the urban CO_2 domes of Los Angeles and San Francisco. At present roughly half of California's petroleum comes from within the state's borders, and much of the rest comes from Alaskan deposits. Therefore each humble splash of ketchup carries within it the atoms of algae that glistened in earlier versions of the Pacific and Arctic Oceans and then slumbered underground for millions of years. After you swallow that dollop, you will carry its carbon atoms around inside your body for a while, then set them loose again into the sea of air that they have wandered many times before.

The atomic ties between you and the earth can be traced through the menus of fast-food chains in ways that scientists have recently developed with the help of stable isotopes, sometimes even using the iconic Big Mac as a new representative of the planet's more ancient ecological food chains. What they are finding runs counter to what most of us might expect under conditions that increasingly homogenize our cultures.

To begin to see how the carbon atoms in a burger can hitch your body to the rest of the world, imagine yourself traveling seventy million years into the past on what is now the South Carolina coast. There, beneath the waters of a warm, shallow sea, prepare to meet a ten-armed creature whose carbon atoms have important things to reveal about your own atomic nature today.

A belemnite is—or was—much like a small squid. But if you were to hold one in your hand today it would look more like a large bullet. Belemnites have been extinct for millions of years, having died out along with the dinosaurs at the end of the Mesozoic, and most of what remains of them are the cylindrical rods of carbon-bearing calcite or aragonite that stiffened their bodies. These endoskeletons look lovely when the limestone surrounding them is chipped away to enhance their streamlined forms, and you may have seen them displayed thus on walls and bookshelves. But your carbon atoms have a special connection to one community of belemnites whose fossils have been used as an isotopic standard against which scientists have measured the effects of diet and pollution on people and ecosystems.

For decades belemnites from a marine deposit in Pee Dee, South Carolina, were crushed, cleaned, and analyzed for their carbon content, which includes a 1.1 percent fraction of heavy carbon-13 isotopes among the normal carbon-12 atoms in the limestone. So popular were the Pee Dee belemnites among geochemists that the supply ran out and another standard was adopted in 1994. Now an updated VPDB (Vienna Pee Dee Belemnite) benchmark serves as an isotopic reference point for carbon-based objects around the world, including humans.

If you were to submit a snip of your hair to an isotope lab for analysis, you would find fewer carbon-13 atoms in it than expected if you had lived before the Industrial Revolution. The difference reflects the ways in which we power our civilization. As with plants and plankton of today, the carbon-13 fraction of the photosynthetic carbon in coal, oil, and gas is proportionally lower than that of the air and water it came from, and fossil fuel exhaust has now lowered the carbon-13 content of all life on Earth by shifting the isotopic balance of the atmosphere and oceans. The idea of carrying carbon pollution inside of you is not just a theoretical abstraction: You can demonstrate it by comparing your skewed isotopic composition to that of a Mesozoic squid.

A chemist would present the carbon-13 content of your hair as a ratio of heavy-to-normal carbon, represented as "$\delta^{13}C$" and pronounced "delta C-13." This value can then be adjusted relative to the belemnite standard, with negative values indicating carbon-13 contents lower than those from Pee Dee. You might thus expect to find your $\delta^{13}C$ ranging between −15‰ and −25‰ (tenths of a percent), but without actual data on hand you couldn't predict the numbers more precisely than that without knowing where you live and what you eat. And this, believe it or not, brings us back to McDonald's.

Writing in the journal *Food Chemistry* in 2011, a research team headed by the ecologist Luiz Martinelli challenged the notion that fast-food outlets have completely homogenized modern Western diets. To do this they examined the isotopic composition of a globally available food, using hamburgers to reconstruct the atomic linkages between people and the original sources of their atoms.

"For the first time in human history," the article states, "consumers can buy . . . the same Big Mac meal from over 32,000 locations in approximately 120 countries." This situation is often portrayed as a globalization of diets, but Martinelli's team pointed out that although the recipes may be consistent worldwide, the actual ingredients still reflect the local isotopic settings from which they were harvested. They referred to this blending of global and local features as "glocalization."

To understand how glocalization works, it helps to remember where the atoms in food come from. The ground beef in a Big Mac emerges from large batches of steer carcasses that can be either domestic or imported. For example, most of the beef in a McDonald's restaurant in New York or Chicago was probably raised in the United States, but most of the meat in the Amsterdam outlets originates in other nations. This means that although you get "two all-beef patties" wherever you go, the beef is not as globally uniform as the recipe is.

The isotopic composition of a steer depends upon the composition of the plants it eats, and different plants carry different carbon signatures depending upon their species and the conditions under which they grow. Maize and sugarcane, for example, have greater $\delta^{13}C$ values ($-10‰$ to $-15‰$) than wheat and soybeans ($-20‰$ to $-35‰$) for genetic reasons and also because plants that grow in relatively dry or hot habitats, such as the Great Plains or Gulf Coast, tend to have higher values.

When Martinelli and his colleagues measured the $\delta^{13}C$ of patties in twenty-six countries, they found clear evidence of the sources of those steer atoms. McDonald's beef in Brazil and the United States yielded values much higher than the readings from Scotland and Austria. This is largely because the Brazilian cattle were raised on tropical grasses and the American animals fattened on maize, and both kinds of plant contain more carbon-13 than the hay or soybeans that the Scottish and Austrian animals ate. Japanese burgers yielded values close to those of Brazil, which might seem unexpectedly low because of Japan's relatively moist, temperate climate. But Japan also imports most of its beef from Australia, where warmer, drier conditions often prevail. Although a Big Mac looks and tastes much the same the world over, its carbon atoms show it to be far more local—or glocal—than it appears to be.

Not surprisingly, the atomic connections between diet, location, and living bodies extend to you, too.

A group of scientists from Utah and Indiana recently followed up on the Martinelli study to examine the effects of glocal food on people from Europe and the United States. By gathering hair clippings from barbershop floors and volunteers in fourteen countries, they found that isotopic signals still link people to their home territories despite the spread of fast-food outlets and supermarkets. In a paper that appeared in *PLOS ONE* in 2012, these researchers showed that the hair of meat-loving Americans tends to carry more carbon-13 than that of Europeans, in part because cattle raised in the United States tend to eat lots of maize. Citizens of Portugal also yielded higher $\delta^{13}C$ values than Finnish and Swiss hair donors, most likely because they ate more seafood and consumed the carbon of vegetables and grains that grew in warm, dry Mediterranean climates.

The authors concluded that their data "capture a geographically structured heterogeneous diet among the sampled countries" that persists beneath a veneer of uniformity. Although most of us no longer construct and maintain our bodies with the atoms of plants and animals that live among or near us, the isotopic signatures of food chains still associate us with specific regions through the organic matter from which we are made.

The atomic manifestations of recent changes in human diets also reach beyond our species to animals that live among us. Writing in the *Journal of Mammalogy* in 2010, scientists from the University of Wyoming and elsewhere described an isotopic shift in a rare, endangered species of kit fox that haunts the alleys and backyards of Bakersfield, California.

Until recently wildlife biologists relied most heavily upon the visible contents of fox droppings as they sought to learn what the animals were eating. In the wilder portions of the San Joaquin Valley, for instance, droppings contained distinctive remains of rodents, birds, and insects, but except for the occasional plastic wrapper there were few identifiable items in the urban scats. The soft, highly processed foods that most Californians discard today leave few traces after passing through the guts of a kit fox, but their carbon isotopes clearly document the switch from wild to supermarket foods. The droppings of foxes from Bakersfield now carry more carbon-13 (higher $\delta^{13}C$) than those of their country cousins.

As the kit foxes of Bakersfield moved into the city, they not only expanded their dangerously shrunken home range: They also became more isotopically similar to the people around them by eating the same glocal foods. Plants in the San Joaquin Valley generally have low $\delta^{13}C$ values, and so do many of the small prey whose carbon atoms are recycled in the bodies of rural foxes. But the urban food chains of Bakersfield are mostly anchored in distant maize and cane fields where $\delta^{13}C$ values are higher, and the difference now shows up in the bushy tails and twitching whiskers of city foxes.

Fortunately, most concerns about feeding wildlife on human "junk food" appear to be unfounded in this case. The Bakersfield foxes reproduce more successfully than do their rural relatives, and their survival rates are higher on average despite encounters with cars and rodenticides. But another kind of shift in the carbon isotope balances in foxes and people is more troubling, and this one is truly global.

All over the world $\delta^{13}C$ values are decreasing from the combustion of fossil fuels. The signs of it are everywhere, in the annual growth rings of trees, in the layered sediments of lakes, in the finely banded limestone of coral formations, and also in your body. If you were to analyze the carbon-13 content of a lock of your hair from your early childhood, it would be higher than what you carry around with you now—assuming that you still have hair and that your diet has not changed tremendously.

Carbon dioxide molecules that are burdened with heavy carbon-13 have extra difficulty in entering the cells of plants and plankton, so forests and ocean sediments from the distant past carried less of it than the atmosphere above them. As we burn those fossil deposits, we flood today's atmosphere with their lightweight carbon, and the resultant dilution of the atmospheric carbon reservoir works its way through the food webs that sustain us.

So, too, do other elements that also emerge from burning coal. In 2012 researchers from the University of Michigan used distinctive isotopic ratios to trace toxic mercury atoms back to the coal-fired power plant that emitted them near Crystal River, Florida. "This study represents the first use of mercury isotope ratios to investigate the near-source mercury deposition resulting from coal combustion," one investigator said in an interview for the University of Michigan Press Service. According to the lead author, Laura Sherman, "This allows us to directly fingerprint and track the mercury that's coming from a power plant, going into a local lake, and potentially impacting the fish that people are eating."

There is no denying it: The very atoms of our bodies confirm that we have more influence on the global carbon cycle than volcanoes. Delta C-13 values in carbon dioxide measured at Mauna Loa, Hawaii, have fallen from an average of about −7.5‰ during the mid-1970s to roughly −8.3‰ at the time of this writing, continuing a trend that began with the Industrial Revolution. Every step down the $\delta^{13}C$ slope represents a boost in the carbon dioxide content of the atmosphere. This global isotopic trend is just one of many reminders that we are still as intimately linked to the earth by atomic chains in this modern age as our earliest ancestors ever were. It will be our challenge to live both well and responsibly within their unavoidable constraints.

5

Tears from the Earth

*A man is an aqueous salty system in a medium in which
there is but little water and most of that poor in salts.*
—*John Z. Young*

The cure for anything is salt water: sweat, tears, or the sea.
—*Karen Blixen (Isak Dinesen)*

The entomologist Hans Bänziger is, to say the least, passionate about insects. He is lucky, too, in that he has been able to make a career following his passion in the forests of Southeast Asia. But Dr. Bänziger is not only lucky; he is renowned among his peers for going beyond the call of duty in his research by offering himself as bait.

In his landmark paper published in 1992 and titled "Remarkable New Cases of Moths Drinking Tears in Thailand," the casual reader is given little warning of what is to come farther down in the dry scientific text. The opening sentence merely states that "10 cases of lachryphagous moths settling at human eyes were witnessed." But when Hans Bänziger tells you that such behavior was seen, he means that it was observed from the closest possible viewpoint, giving new meaning to the term "eyewitness."

Many insects land on larger creatures or objects in order to lick salty fluids from them. A species of sweat bee that was recently discovered in Brooklyn, New York (and appropriately named "*L. gotham*"), occasionally pesters city residents by landing on glistening arms, legs, and faces in hot weather. Colorful butterflies cluster like flower petals around the muddy rims of roadside puddles in order to lap salt and other substances. But eyes? Although tears certainly are salty, most of us consider our eyes to be no-go zones when it comes to insects.

Not, however, for one dedicated entomologist.

In an entry from November 1989, Bänziger described watching moths hover near some cattle in a forest clearing. It was just after sundown, and a half-moon was visible. One of the moths flitted over to him, landed on his wrist, and began to sweep its threadlike proboscis gently over his skin. Minutes later it landed on his bare leg where it continued to drink from the thin film of sweat. Then it moved to his cheek, and then to the lower edge of his right eye. "The perception I now felt was . . . comparable to that of a particularly edgy grain of sand being rubbed between eye and lid."

At this point most of us would have swiped the pesky visitor away. Instead, Bänziger pointed a camera at his face and snapped a flash photo, which appears in the article as well as in this book.

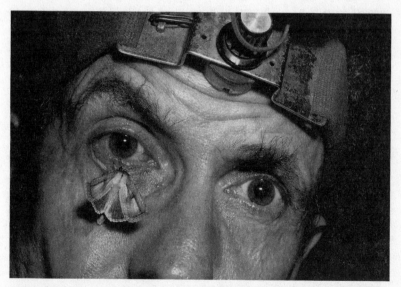

A tear-drinking moth meets Hans Bänziger in Thailand. *Photo, amazingly, by Hans Bänziger himself*

The irritation was apparently coming from the moth's tiny clawed feet as they scrabbled at his inner eyelid: Later investigations suggested that this was intended to stimulate the flow of more tears. In some species, the threadlike mouthparts are rough-edged enough to cause the requisite irritation, but this moth's delicate proboscis could barely be felt as it swept back and forth over the smooth, curved surface of his eye.

"Unfortunately," he reported with no apparent irony, "the flash scared the moth away . . ."

Nearly a year later Bänziger was again at work in the forest at night. Suddenly he noticed a dark object hovering in front of his face and felt something "like a fine straw" sliding between his lips. The moth then probed his nostrils, causing "almost unbearable tickling," after which it rose higher to lightly sponge moisture from his eye.

This species was gentler in its approach, and Bänziger described the sensation as "surprisingly mild," although he also admitted with what must have been no little understatement that having "one of the largest tear-drinkers hovering in front of the face with a 4-cm long proboscis aimed straight into the eye is somewhat trying." The encounter ended abruptly when he lowered a collecting net over both his head and the moth.

More trying experiences occurred under a full moon in a pasture near a rural village. A moth that had been flying around several mules and horses "suddenly turned its attention to me." This one followed him for a hundred yards before settling to drink from his left eye. "I closed the eye, pressing the lids tightly together, but the moth would not leave." Another visitor on that same night was even more aggressive. "I felt pain due to clawing of the lid," he reported. After two minutes of this, even Bänziger had had enough: "After having indulged in this tear-letting for the second time that night, I could not bear it any longer and caught the tormentor."

It isn't flattering to think of yourself as being a walking mud puddle to an insect, but your eyes and puddles do have much in common. Both contain water, and both are rich in biologically useful atoms that come from the rocks of the earth's crust. In many ways your eyes are much like the mineral springs and salt licks that have attracted wildlife for millions of years, and you share a taste for salt with many other animals.

Of the five officially recognized tastes in the human mouth—sweet, bitter, pungent umami, sour, and salt—only one is aimed specifically at a single mineral element. The first three arise from relatively complex carbon compounds, and sourness arises from acidity produced by hydrogen ions. Only saltiness stems from an atom that comes to you from rocks and soils: sodium, which carries eleven protons and twelve neutrons in its nucleus and comprises half of the atoms in table salt. Sodium puts the tang in your sweat and blood, and it also helps you to think, move, and perceive the world around you. Without enough of it in your body fluids, most of your cells would swell up and die. With too much of it, they would shrivel into microscopic prunes. It makes nearly 98 percent of all liquid water on Earth undrinkable to you, but in doing so it also helps to determine what lives where on this ocean-dominated planet.

You have probably heard that most of us eat too much sodium in comparison to what our ancestors used to consume. This may or may not be true. Some historians point out that heavy salt use for food preservation dates back millennia, that salt mining was common in China four thousand years ago, and that Roman soldiers were once paid with salt, hence the term "salary." In any case billions of dollars have been spent in developing and marketing foods that cater to our desire for salt, while medical professionals warn of associations between sodium and hypertension, heart disease, and other ailments. But it would also be a mistake to carry too little sodium inside you, for reasons that are both interesting and important to your health.

Technically speaking, sodium is only half of the atomic basis of saltiness. Normally a positively charged cation such as sodium travels in the company of a negatively charged anion of some sort, and in table salt and the brine in your tears the primary partner is chloride, an ionized version of the element chlorine, which is also the active ingredient in household bleach. But although other ions also play important roles within your body, sodium remains the star for several reasons. For one thing, you can taste it. And until today's Western diet evolved around endless supplies of processed salt, sodium was harder to come by than another important dietary cation, potassium. Vegetables are full of potassium because plants use it for maintaining moisture balances in their cells and for operating the breathing holes in their leaves. Sodium, in contrast, is rare in most plants, simply because they don't use it as we do. An apple, for instance, can contain a hundred times more potassium than sodium, and the imbalance in a mango or nectarine can be twice as large.

Because of the scarcity of botanical sodium, many animals seek out sources other than plants, often with the aid of sodium-sensitive taste cells. Whether you get yours from a shaker or a bloody red steak, you run your body on a relatively rare commodity in the food webs of the world. For many herbivores there is no such thing as too much sodium in the diet, and if the vegetable kingdom won't supply enough of it they sometimes turn to geological sources.

There are many possible reasons why salt licks draw wildlife, but there has been surprisingly little definitive research in this regard. For many years scientists simply assumed that wild game was licking mud in order to stock up on sodium without bothering to analyze the mud itself. More recent work has shown that many so-called salt licks have surprisingly little salt in them, and that other substances may occasionally be the target instead.

South American parrots, for example, are now thought to eat river-bank clay simply because it helps to detoxify harmful chemicals in the seeds that they consume. Amazonian fruit bats visit clay deposits more often than insect-eating bats even though their diets are richer in sodium, again suggesting a digestive function. And a study at Ngorongoro Crater, Tanzania, recently concluded that salt may be a cue that helps wildlife to find even rarer elements such as cobalt or selenium, which can't be tasted but are also important for nutrition.

Many people indulge in geophagy (earth eating) as well, though not so much for salt as for other reasons. A fascinating overview of this behavior that was compiled by the essayist Beth Ann Fennelly shows that more people eat clay than you might suppose. Although reluctance to admit to dining on dirt can make reliable numbers difficult to obtain, studies in rural regions of the southern United States have found that up to half of all women surveyed consume clay during pregnancy, and studies in Africa find rates as high as 90 percent. Most cases are associated with pregnancy but many men indulge, too, suggesting both nutritional and digestive needs.

In the Andes Mountains, potatoes are sometimes dipped in a slurry of wet clay as though it were gravy, and clay is often sold in marketplaces alongside the produce in Peru. Presumably this protects against the harmful effects of botanical toxins that can occur in potatoes, which belong to the same plant family as deadly nightshade. Clay can also be a famine food that fills empty bellies but unfortunately can also cause intestinal blockages and introduce parasites. And other studies suggest that people eat dirt simply because they like it.

Fennelly described a recent soil-tasting event in a San Francisco art gallery that was modeled on a high-end wine tasting. Soils were mixed with water in wine goblets to release aromas for the attendees to sniff and savor. According to Fennelly's sources creamy white kaolinite clay "tastes like rain with a hint of peanuts and it melts in your mouth like chocolate." One family in Mississippi would regularly "fry it and eat it warm," and the author likened her own taste testing of Georgia kaolinite to gnawing on a chocolate Easter bunny. And if you ever consumed Kaopectate as a digestive aid before a recent reformulation removed the clay for which it was named, then you, too, are a geophage.

However, most of these examples do not really amount to dirt-eating in the strictest sense. When you ingest clay you don't use its atoms to build or operate your own body as you would with a pinch of salt. Clay is indigestible, being made of minuscule flakes of glassy minerals, and its role in your digestive system is that of a transient sponge. Undesirable molecules found in some plant tissues tend to stick to the surfaces of the flakes, which renders them less harmful and easier to excrete.

The atoms in salt, on the other hand, quickly become important parts of you, and the licking behavior of butterflies and moths is a clear case of sodium hunting. In fact it has been investigated in such detail that we can now deduce with some confidence what happened to the sodium atoms that Hans Bänziger donated to the tear-drinking moths of Thailand. If related studies are a reliable guide, then Bänziger's tear sodiums became wedding gifts to lady moths.

Although the moths of New York State don't seem to care much for tears, many of them do drink from mud puddles at night as butterflies do during the day, and a study by the Cornell University scientists Scott Smedley and Thomas Eisner demonstrated that at least some of them gather large amounts of sodium in doing so. Smedley and Eisner found that most of the puddling by *Gluphisia septentrionis*, a rather plain-looking gray, tan, or cream-colored moth, is done by males, and that they spend many hours drinking at the local mud bar along with other males. A *Gluphisia* moth can pump five hundred times its body weight in mineral-rich water through its body at one sitting, squirting jets of filtered fluid through its backside over distances of more than a foot.

If you were to do such a thing yourself, however odd that might be, a single sitting would have you jet about nine thousand gallons of your own watery waste behind you in graceful arcs some two hundred feet long and you would also nearly double the sodium content of your body. Like you, however, moths don't normally need so much sodium—unless they're planning to share it, which is exactly what *Gluphisia* moths do.

In the Cornell study, pairs of male and female moths were allowed to mate in cages under controlled conditions. Females who mated with males who had recently puddled contained much more sodium than females who had consorted with nonpuddlers, and after mating, the puddling males contained less than two-thirds of the sodium that they had before. Sodium was as much a part of this lepidopteran lovemaking as the gametes, a nuptial gift akin to a suitor's box of chocolates. But what would a female moth do with so much sodium?

The answer became clearer when the investigators studied the eggs of the moths. Eggs from salt-gifted mothers carried much more sodium than the others, showing that each courted female relinquished a third to half of her nuptial winnings to her young along with her genes. In other words, sodium atoms from puddles moved directly from father to mother to children. In contrast, females who did not mate with puddlers had to give up as much as 80 percent of their personal sodium supplies in order to properly stock their eggs.

What does this have to do with you and your own relationship to sodium?

For one thing, it shows that there is something special about this element as a commodity in the animal kingdom to which you also belong. It shows that most plant foods in your diet are a poor source of it, because they are built differently from animals. And finally, it shows that we are all so full of sodium atoms that insects flock to us in order to lick us. The sweat on our skin and the puddles in our eyes have close to one-quarter of the salinity of seawater.

All of your sodium atoms came from minerals in the earth's crust. There is no escaping it: You are an earth eater, and you also consume organisms that are themselves geophages. Among your early human ancestors, most dietary sodium came directly from food, but today most of us also shake, spoon, and drizzle it directly onto our meals or have it done for us by those who produce and package our food. Table salt is an edible version of mineral sand that dissolves in your mouth, stimulates your taste buds, enters your bloodstream, and sometimes emerges from the pools of your eyes in times of joy and sorrow.

The stories of how sodium atoms end up in your tears could lead your imagination from fiery magma to high mountains, from deep mines to sun-baked plains of gleaming crystals, or from plants and animals to your dinner table. But at some point along the way, most of the sodium in your body spent millions of years adrift in oceans of the distant past.

Until shortly after the American Civil War, nearly all the commercial salt sold in the United States came from a region of central New York that once was a shallow sea. Between five hundred and four hundred million years ago, downwarping of the land allowed an arm of the ocean to fill a geological depression near the site of today's Syracuse. Millions of years later the isolated sea dried out, leaving vast sheets of salt between blankets of mud that stretched as far west as today's Ohio. Over the ages some of the salts dissolved into local groundwater and seeped upward through hundreds of feet of sedimentary rock to reach the surface as brine. In recent centuries indigenous people collected the oozings of marshy springs along the shores of Onondaga Lake, boiled or sun-dried the water away, and harvested the salt.

By the nineteenth century, industrial-scale exploitation of the salt deposits and springs gave Syracuse the nickname "Salt City," and the oldest evaporation plant in the country still operates at nearby Silver Springs, now under the ownership of the Morton Salt Company. If you use Morton salt at your table, then there is a good chance that those tiny cubic crystals that famously pour even when it rains (thanks to an inert powder that keeps the grains from sticking together) came from Paleozoic ocean deposits beneath Silver Springs. But what were they doing in that old seawater in the first place?

Sodium clings to chlorine when seawater evaporates into rock salt, but these elements emerged from very different sources before they ever met one another in an ocean. Most of your chlorine atoms came from volcanic fumes and super-heated fluids, and your sodium was originally weathered out of solidified lava or magma.

When rain and groundwater degrade rocks, they extract sodium atoms and sweep them downhill where they may eventually reach the ocean or an enclosed basin such as Israel's Dead Sea or Utah's Great Salt Lake. They go willingly on that journey because water molecules find them very attractive. If you could shrink down small enough to watch the atoms of a basalt boulder as it crumbles ever so slightly during a rainstorm, you could see new cohorts of sodium atoms begin the same journeys that most of the sodium in your tears, sweat, and blood once made. And while you're at it, why not also travel four hundred million years back in time to do this on the eroding flanks of a hillside in what is now upstate New York? If you do, then you might see exactly where many of your personal sodiums came from.

Each sodium atom in the weakening framework of a feldspar grain trembles in its fetters as though eager to join the water molecules that are tearing at the mineral surface. A water molecule carries slight charges, with the paired hydrogens on one end of it being slightly positive and the central oxygen being slightly negative. Because sodium atoms carry a positive charge in solution, the negative ends of water molecules coax them away from their mineral lattices through the attraction of opposite charges. Like prisoners released by a liberating army, your sodiums are pulled free after millions of years of bondage.

Half a dozen water molecules close in around each sodium cation, pressing their negative ends against it. Around that "first hydration shell," a second layer of water molecules packs in as well, inflating the virtual size of the cation to several times its original volume. The ball of sodium and water then tumbles away into the general flow of runoff, along with other escapees from the same rock face.

Joining a stream that drains into the warm, shallow sea to the west, each sodium ion remains at the center of a huddle of water molecules as it wanders among ancestral sharks and trilobites whose fossils still lie entombed in stone. An ion's watery companions are not reliably loyal, however. Any given water molecule soon skips away, opening a space for another one to replace it. But if the supply of replacements runs dry, so to speak, then sodium may find itself bereft. This is what happened when slow Earth movements isolated Paleozoic New York from the main ocean, causing the inland sea to dry out.

In the increasingly concentrated brine, other dissolved substances faced the same abandonment by their evaporating chaperones. The most numerous of these were chloride anions, which became increasingly attractive partners for increasingly desperate sodiums. Like sodium, these chlorides resembled popular celebrities while in solution, although their watery fans pressed positive rather than negative ends against them. One might imagine that this popularity would be even more surprising to chlorine than to sodium, because chlorine atoms have more trouble fitting into the close-knit atomic communities of mineral grains than sodium does.

Unlike sodium, chlorine is rarely locked into the crystal lattices of most minerals. Although you can find small quantities of it in igneous rocks such as basalt or granite, it usually resides in tiny pockets of fluid that are trapped within or between the mineral grains. Because of this, geologists sometimes measure the chlorine content of seemingly impermeable gems such as tourmaline by boiling them in water and then letting the liquid evaporate away.

Chlorine atoms usually escape from magma as hydrogen chloride, the same corrosive acid that your stomach contains. Most of it drifts off in wisps of gas, and geologists estimate that volcanoes release several million tons of hydrogen chloride into the atmosphere every year. It then dissolves into raindrops and eventually ends up in the oceans.

According to a report by Herbert Swenson of the U.S. Geological Survey, a cubic foot of typical seawater contains 2.2 pounds of salt, more than two hundred times more than a similar volume of typical lake water. About 85 percent of the dissolved material is sodium and chloride ions, with magnesium and sulfate ions making up most of the rest. Although rivers deliver four billion tons of these solutes to the oceans every year, the global average salinity of seawater remains fairly stable because similar quantities of salts are buried in sediments on the seafloor.

Today the oceans contain about fifty million billion tons of salt, enough to blanket the continents with a layer of snowy crystals five hundred feet deep. Something much like this actually happened, though on a smaller scale, in central New York about four hundred million years ago.

In the shrinking sea, sodium and chloride ions drew ever closer in the thickening broth of mineral elements. When the last waters evaporated, sodium chloride crystals piled up in sheets of rock salt dozens of feet thick. Over the ages the salt beds were buried under more than a thousand feet of shale, limestone, and other sedimentary deposits. It is refugees from that petrified ocean that escape into the evaporation chambers of the Morton Salt Company today, and that eventually make it to your table.

So as you shake that next dash of salt onto your food, think of the long-standing relationships that you are about to unravel. After long ages apart, ionic opposites were wed in a dying sea. When you toss them back into solution within your body the jostling of watery crowds will separate them again, but don't weep for those ancient ions. They are about to begin one of the most interesting chapters of their existence inside the living sea of atoms that is you.

What happens to sodium and chlorine when you swallow them? For starters, don't swallow either of them in pure elemental form. Purified sodium explodes on contact with water, and chlorine gas can destroy your lungs as it did to soldiers on the battlefields of World War I. You want them tamed in ionic form, which is easy to do when you encounter them in salt because the water in your saliva rapidly dismantles the crystals into their charged components.

One of the first things that sodium does upon dissolving in your mouth is to tell you that it has arrived. Tiny pores in thousands of taste buds on your tongue welcome some of the cations inside, where they trigger nerve impulses that your brain interprets as the taste of salt. Although the physiology of taste remains somewhat mysterious for now, sodium channels such as these are well known not only as portals to flavor, but also to the very act of sensation itself. More on that shortly.

Contrary to what you may have learned in school, taste is not restricted to discrete patches on your tongue. Most of us can actually taste salt all over the tongue as well as on the walls of the mouth, and although certain cells are especially sensitive to specific tastes, many of them can respond to more than one stimulus. Recent research shows that this is especially true when it comes to sodium, and for good reasons.

You can't store sodium as you do the calcium in your bones, so your body has to monitor and regulate it carefully in dissolved form. Your kidneys, colon, and sweat glands manage most of the losses, while taste cells in your mouth help your brain to control how much sodium enters your body when you eat or drink. When sodium concentrations in a mouthful of food are low, the salt specialists among your taste buds urge you to eat more and enjoy. But if a meal becomes briny enough to pose an immediate risk of overindulgence, it also triggers other taste cells that normally respond most strongly to bitter substances.

Such interactions among salty and bitter sensors may also help to explain why low-sodium salt substitutes, such as potassium chloride, often deliver a bitter aftertaste that can reduce incentives to use them. The only other atom that triggers salt receptors as pleasantly as sodium is lithium, but unfortunately you can't improve your health by coating your potato chips with lithium chloride. The latter could poison you if not properly administered.

The presence of taste sensors in your mouth hints at the importance of sodium to your body. But what, exactly, do you need it for? This can be answered by showing what happens when someone is deprived of it.

Noting in the *Proceedings of the Royal Society of London* that "no papers of any value" had yet been published on salt deficiency in humans, the King's College Hospital physician Robert McCance made what he called a "direct experimental attack" on the topic during the 1930s by putting volunteers on a salt-free diet and subjecting them to daily bouts of heavy sweating.

The sweating was done while each subject lay upon a mattress inside a makeshift cylinder that was equipped with heat lamps, as an assistant swabbed up the drippings to measure the salt losses. After an uncomfortable four-and-a-half-hour trial, one subject reported losing more than half a gallon of body water. He wrote in his journal,

"Out, feeling rotten few minutes, washed down. Then OK. Total loss of weight . . . 2150 gm."

Cooking for themselves in their homes, the volunteers ate only bread, synthetic milk, thrice-boiled vegetables, and other sodium-free foods. All liquid and solid secretions were collected, dried, and weighed for analysis.

Signs of trouble began within days as sodium levels in body fluids plummeted. "As the deficiency developed all three subjects lost weight," McCance noted. "Their cheeks fell in and they began to look ill." All sense of flavor faded, which helps to explain why it is so difficult for many people to adjust to low-sodium diets. Cigarettes lost their taste, and fried onions produced only "greasy sweetness which was extremely nauseating." Breathlessness became a problem, and energy levels fell so low that one man's arm "got tired shaving" and his jaw "got tired eating toast."

After ten days, sodium concentrations in sweat were less than a third of what they had been on day one, and the levels in urine fell to virtually nil. Blood samples became dark and viscous, but blood pressure, urine volumes, and pulse rates remained normal because the subjects drank water freely. This was not dehydration but a slow decline by desalination.

Fortunately the volunteers recovered quickly when the experiment ended, and no permanent harm was done. Their sense of flavor returned within minutes of eating salty bread, butter, and eggs, and within two days their energy levels were back to normal. One man was so relieved to feel normal again that he "jumped off the bus while it was still going and ran up the stairs."

Dehydration can be even more harmful than running short on sodium, however, because water forms most of your body in addition to providing a medium for chemical processes to operate in. Eating, drinking, and releasing wastes are all part of a continuous balancing act, and you remain properly inflated and functional only by adjusting supply and demand on a regular basis.

A typical, relatively sedentary American adult loses close to thirty ounces (three or four cups) of sweat each day through several million microscopic glands that squeeze fluid out onto the surface of the skin like toothpaste from a tube. A square inch of your forehead can contain hundreds of sweat glands, but other parts of your body have considerably more or fewer. A similar-size patch of skin on your arm, for instance, contains nearly a thousand such glands but the fingerprint ridges of your palms have two or three times as many per square inch. Even the composition of your sweat varies with location and situation. The liquid that cools your legs can be less salty than the dew on your arms, and the odors emerging from your armpits are the work of bacteria that convert oily, protein-enriched "emotional sweat" into aromatic compounds that can act as social cues.

Along with evaporation from the lungs and bulk excretion through the kidneys and bowels, a sedentary adult typically loses five or six pints of water that must be replaced daily in order to maintain a net water supply of roughly eleven gallons. With heavy exercise and a hot climate added to the equation, you can easily lose two to three gallons over the course of a day, along with nearly an ounce of salt. This continual loss of moisture can make you more vulnerable to dehydration than to hunger, and if your water intake is interrupted for long enough the structural effects begin to mimic the dissipation of your body at death.

In 1906 the geologist-anthropologist W. J. McGee delivered a report to a medical conference that was titled "Desert Thirst as Disease." During the summer of the previous year, McGee helped to rescue a man who had wandered for more than a week in the Arizona desert without adequate water. His account of the ordeal became a classic in the medical literature, and it illustrates how important the balance between salt and water is to your physical appearance in addition to your well-being.

A Mexican prospector named Pablo Valencia left McGee's tent camp on horseback with a companion, heading for a remote gold claim. That night the companion returned with both horses, saying that Valencia had decided to continue on foot with only a small canteen of water. After a desperate search, the barely living "wreck of Pablo" was eventually found crumpled on the sand beneath an ironwood tree.

According to McGee, half of all waterless travelers succumb to the Arizona desert within a day and a half of running dry, but Valencia somehow managed to walk, stumble, and crawl more than a hundred miles over eight days and nights. His only moisture came from his canteen, an occasional insect or scorpion, and his own scant and concentrated urine.

During his recovery Valencia described how he progressively jettisoned clothing along with his tools and gold nuggets, and watched vultures come almost close enough for him to touch. He became so dehydrated that deep scratches left by thorns and stones didn't bleed. His nose shrank to half its original length, his lips vanished, and his tongue was "a mere bunch of black integument." Even the tissues surrounding his eyes retreated so far back into their sockets that the whites of his eyeballs were exposed.

Most sources report that we begin to feel severe thirst at the command of cellular sodium receptors after losing only 1 or 2 percent of our body water. Convulsions and delirium generally begin with a 10 to 15 percent loss, and higher deficits are usually fatal. Amazingly, Pablo Valencia lost a quarter of his original weight, mostly through evaporation from his skin and mouth. Nonetheless he was back on his feet within a week of his rescue, and lucky to be alive.

The examples of McCance's volunteers and McGee's miner illustrate some of the ways in which shortages of salt and water can affect you, but the processes that cause those effects are relatively simple. It all comes down to the random thermal motions of atoms.

The lethargy of the sodium-starved subjects, for example, arose from the relative movements of salt and water among cells. Osmosis and diffusion are the net migration of water and dissolved substances, respectively, from an area of high abundance to low. Both are powered by the heat-driven dancing of atoms, and your health depends on rather precarious relationships between these two processes. The delicate membranes of your cells allow water molecules to cross more easily than salt ions, which are fattened by shells of hydrogen-bonded water. Sodium therefore enters and leaves your cells mostly with the aid of protein channels and pumps. In the restless atomic realm, simply placing a selectively permeable membrane barrier between a cell and its surroundings can produce remarkable results.

If you were to place one of your red blood cells into a drop of pure water and watch it under a microscope, you would soon see it expand and then burst like an overinflated balloon. This is because water molecules can easily enter the cell by dancing across its thin membrane, but the salt ions are trapped inside. The imbalance of mobility between incoming water and the confined salts swells the cell. In very salty water, however, your blood cells would shrivel instead: Osmosis of water from the interior would not be matched by inward diffusion because the salt ions cannot cross the membrane barrier.

Such differential movements of water and salt among your cells can have powerful effects. For example, running short of sodium could cause your red blood cells to expand until they no longer pass as easily as they should through your narrow capillaries. The resultant blockages would quickly cause oxygen shortages all over your body, as happened to McCance's lethargic patients. Unpleasant effects of dehydration that McGee described can also arise from changes in the sizes of cells due to osmotic imbalances, and you don't have to wander for days in a desert in order to experience them. You can also trigger them by tipping the scales in the other direction with excessive hydration.

When my fellow college students and I foolishly competed in a water-drinking contest during the 1970s, we never dreamed that it could be dangerous. Our rules were simple: a glass per minute, and "one go, no blow." Within less than an hour we all quit in trembling misery, rushing to the bathrooms to unload and feeling somewhat drunk as we ran. We thought that we shivered and staggered from cold, and that may have been partially correct. But it could also have been the start of something more sinister, which, unknown to us at the time, has ended the lives of more determined competitors.

There are many sad stories about people who have convulsed and died from drinking too much water, usually in some sort of contest. In 2007 a young mother of three perished shortly after drinking too much water for a radio station's "Hold Your Wee for a Wii" contest, and thrill-seeking college students are occasional victims of this, as well. The swelling of brain cells can constrict the flow of blood inside your skull, and if your nerve cells swell too much they begin to malfunction. This is why athletes often drink Gatorade or other ion-rich electrolyte solutions after prolonged heavy sweating rather than glugging pure water, because such drinks better maintain their crucial osmotic balance.

Your tears harness similar processes in ways that help to protect you against infections. Lysozyme, an enzyme in tear fluids, attacks the stiff cell walls of bacteria. When airborne bacteria land in the wet pools of your eyes, their cell walls become so weakened by lysozyme that they can no longer resist osmotic swelling, and that tempting oasis becomes a death trap as they distend and disintegrate. Penicillin and other antibiotics likewise weaken bacterial cell walls, thereby enlisting your body water as a defensive osmotic weapon.

But the diffusion of sodium across your membranes also has more complex effects than the mere distortion of cells. It also produces your thoughts and feelings as well as the countless movements of your body.

As in a computer, the hundreds of billions of threadlike cells in your nervous system generate electrical fields, but they don't transmit electricity as wires do. Cool a copper wire, and it conducts electrons just fine. Cool a nerve cell, or neuron, too much and it stops working properly, as victims of hypothermia can attest. It is not electrons that rush through the networks of your nerves, but disturbances that are more akin to waves on the surface of an ocean than the currents within it. A primary driver of those waves is the thermal dance of sodium ions.

Wiggle your toes now, if you will. Sodium helps you to do this, and it also tells you what it feels like to do so. All such voluntary movements as well as your sensations and thoughts are manifestations of nerve impulses. And all such impulses involve the diffusion of ions across the thin membranes of neurons.

To imagine how this works, think of thousands of sports fans in the bleachers of a stadium as they do "the wave." As soon as one person raises and drops his or her arms, the next person repeats the motion and so on down the line, thereby producing an undulation that ripples through the crowd. None of them leave their seats: It is only the disturbance that travels, and if orchestrated correctly it can move faster than a person could run.

A nerve impulse is much like a stadium wave, but instead of people flapping their arms up and down, vast numbers of tiny channels open and close. A single neuron can carry millions of these channels, which briefly open passageways through the cell membrane that are available only to sodium. Each channel pore is surrounded by proteins that can shear the shell of water molecules away from a sodium cation and open a custom-fitted tunnel just wide enough for it to squeeze through.

When it opens for a fraction of a second before snapping shut again, such a channel allows thousands of cations to diffuse into the neuron, where sodium is roughly ten times less abundant than it is on the outside of the cell. This triggers neighboring channels to do the same until the rush of diffusing sodium reaches the end of the elongated cell. From there the signal passes to other neurons or muscle cells, which can relay the message further or otherwise react to it.

In the case of your recent toe wiggle, the decision to obey my request spread through your brain at nearly a hundred feet per second. This triggered a ripple of sodium diffusion that rolled down your spinal cord at high speed: Impulses such as these can travel at more than two hundred miles per hour. Then it shot down your leg to your toe in another split second. When the requisite muscles twitched and moved your toe, sensory neurons fired sodium waves back up to your brain to report "mission accomplished."

It is all so mechanistic, so strictly elemental in nature, that it is difficult to believe that invisible atomic waves allow my written words to speak to you right now through neurons that connect your eyes to your brain. But they do. Just as flickers of digital information allow a computer to produce a virtual reality so compelling that you might wish to live within it, the serial diffusion of atoms from a salt shaker helps you to see, feel, hear, taste, smell, and think your way into a vivid reality of your own making.

This simplified version of the process illustrates the importance of sodium to your nervous system, although there is more to the story that we need not go into here beyond some brief mentions. Chloride channels help to regulate the charge balance of your neurons, and calcium channels on the downstream ends of neurons help to pass the signal on to adjacent cells. Potassium ions await the arrival of sodium waves before diffusing out of your neurons through floodgates of their own, and the whole system must be reset by tiny molecular pumps that toss sodium back out and potassium back in before another signal can occur. Potassium is also the dominant wave ion in the neurons of your inner ear, so the next time you eat a potassium-rich banana you can reasonably imagine atomic whispers of the tropics emerging from the auditory cells inside in your head.

These events don't take long: Some neurons can recharge and transmit new signals more than five hundred times per second. But they cost you because although osmosis and diffusion are automatic, membrane pumps run on food energy. According to some estimates about one-fifth of the energy budget of your body at rest is spent on powering your nervous system, and every fifth breath of air is drawn to provide oxygen for the production of that energy.

It is fortunate that you don't have to think about all this in order to function. Your atoms handle most of it for you, day and night. And it is fortunate for the rest of animal life on Earth, as well, because all creatures from salt-licking deer to tear-sipping moths use sodium waves for nerve signaling just as we do.

Some animals even use the need for sodium diffusion itself as a defense. Several species of aquatic newt carry a protective nerve-targeting chemical called tetrodotoxin. This is the same deadly agent in fugu, the Japanese puffer fish that is notorious among thrill-seeking diners for its tendency to kill those who eat it when it is prepared incorrectly. Many times more powerful than cyanide, tetrodotoxin blocks the sodium channels on neurons, causing paralysis even in small doses. The toxin needs only to incapacitate a critical number of tiny sodium channels, so just the few milligrams that might fit on the head of a pin are enough to kill an adult within hours.

Some insecticides, such as permethrin, ruin the nervous systems of insects by overstimulating their sodium channels instead. Permethrin binds to the sodium channels of insects and holds them open so the neurons can't be reset properly, thereby leaving pests writhing, paralyzed, or dead. Your own neurons are sufficiently different from those of insects that permethrin doesn't interfere with them, but if you dislike being around neurotoxins on principle then you might also want to avoid chrysanthemums, which many people use to no obvious ill effect in salads. They produce a similar nerve-abusing insecticide, called pyrethrin.

Many other plants also make defensive neurochemicals of their own, from the nicotine in tobacco to the distinctive flavor molecules in spices. The apparent heat of a relatively mild jalapeño pepper is a sensory illusion that is triggered in your mouth by capsaicin, a molecule that tricks your sodium channels into initiating signals of heat and pain. Bite a fiery habañero and a howling chorus of deluded neurons in your mouth will fling open their sodium gates as they would if you had swallowed glowing coals. So convincing is the illusion of warmth that your sweat glands can even be fooled into drenching your skin in response.

The next time you indulge in a hot spicy meal, you might begin to think differently about the salty water that glistens on your skin and wells up in your eyes. These are not just your own sweat and tears, but also components of the earth itself.

Your sweat and your tears are salty because they come from briny lymph that seeps from your blood vessels like mineral-rich groundwater. You produce tear fluids continuously from lacrimal glands in the undersides of your eyelids, and other glands also add evaporation-resistant films of oil to the surfaces of your eyes, mucus to the basal layers of liquid, and germ-killing lysozyme throughout. These substances moisten and protect your cornea as the blinking of your lids spreads them smoothly, and then they soak back into the deeper recesses of your body through pores in the inner corners of your eyes.

The osmotic balance of your tears is similar to that of your blood and lymph, but your lacrimal glands provide them with slightly less sodium, which presumably reduces losses to the environment. By opening or closing various membrane channels and selectively pumping salt ions here and there, the glands encourage water to follow by osmosis into canals that empty like salt springs into a "lacrimal lake" in the lower sectors of your eyes. Each spring-fed lake normally holds up to thirty microliters of liquid, close to the volume of a typical raindrop, but an irritant, a deep yawn, or the neuronal signals of joy, pain, or sadness can make it overflow into tears.

Imagine the journeys that the sodium atoms in your tears have taken before rolling down your cheeks, perhaps to trigger the salt sensors of your tongue or the lacrimal glands of a sympathetic friend. Locked away in stone for long ages, these same atoms might once have wandered among great forests of coral and splashed the sands of pristine tropical beaches. Mingling with them are atoms of chlorine that once hissed in volcanic vents, floated in the same primeval seas, and met their ionic partners there before tumbling into your mouth from a tasty morsel of food.

And if you prefer to keep those well-traveled atoms with you while dining outdoors on a warm evening in Thailand, then you might want to keep an eye out for moths, as well.

6

Life, Death, and Bread from the Air

It is through the laboratory that starvation may ultimately be turned into plenty . . . The fixation of atmospheric nitrogen is one of the great discoveries, awaiting the genius of chemists. —Sir William Crookes

There is no evil in the atom; only in men's souls. —Adlai Stevenson

Atoms are what bombs and toxins are made of, and many people fear them because of it. But the most wonderful things in life are also made of atoms, as are you and I. Bees can make delicious honey although they can also sting, and people may seem to be satanic or saintly when in fact both traits blend within all of us. This multifaceted nature of things is nowhere more vividly personified than in the life of a German chemist named Fritz Haber.

The inventor of a way to create explosives and fertilizer from nitrogen in the air, Haber contributed to the deaths of millions, possibly including the suicides of his wife and son, but he also contributed to the sustenance of billions over the course of the last century. For his invention and its industrial applications he shared a Nobel Prize in 1918, an award whose endowment, ironically, was based on a fortune made by Alfred Nobel through the production of nitrogen-based explosives.

The legacy of Fritz Haber revolves around nitrogen, a vital atomic element that shares his multifaceted nature. It comprises more than three-quarters of every breath you draw, every sound you hear, and every word you speak. Sometimes it helps to acidify lakes or reduces buildings to rubble. Through the ages it has linked people, wildlife, and plants to dirt-dwelling bacteria, thunderbolts, and the iconic blue of the sky. And, perhaps most astoundingly, it now links much of the atomic framework of your body to Fritz Haber.

Look at the sky on a clear sunny day, as you have surely done many times before.

Now really look at it.

How strange it is. At first glance, you might think that the sky is a solid ceiling as the ancients did. When the book of Genesis describes the "firmament" of the heavens, it really means "firm." But of course there is nothing so solid about that blue sea of air. What you're seeing is a luminous haze, a diffuse and permeable mist that encases the illuminated half of the earth like a fogged contact lens. The sun shines through it, meteors penetrate it without leaving holes, and when a bird, cloud, or jet passes you see it clearly in the transparent medium overhead. So where does the blueness come from?

This question probably dates back to the earliest days of language, but the modern scientific answer to it crystallized only recently around the work of many investigators including another Nobel laureate, Albert Einstein. Known as both a peace activist and as the father of the atomic bomb, Einstein made numerous discoveries that are less widely known than his research on relativity theory. In 1910, while his friend Haber was working on nitrogen as a source of ammonia, Einstein published a paper that helped to reveal nitrogen's role in creating the color of the sky.

To understand that role, it is important to remember that nitrogen comprises 78 percent of the gas in a typical parcel of air. As with the oxygen gas that makes up most of the remainder, these molecules are paired atoms of the same name, but the twin atoms in a nitrogen gas molecule—or dinitrogen—look and act differently from the other components of the atmosphere.

When you draw your next breath, some dinitrogen will diffuse into your body along with the oxygen. If you are a scuba diver, you know that breathing pressurized air under water puts extra nitrogen gas into your bloodstream, which can bubble out of solution if you surface too rapidly without decompressing. Such bubbles can cause severe joint pain, block circulation, and damage nerves in a potentially fatal condition known as "the bends." But under most circumstances your cells ignore nitrogen molecules as irrelevant transients. Although you need nitrogen atoms to build and maintain your body, you can't simply inhale your nitrogen supply as you do your oxygen. You have to eat it, and to do that you need the help of other living things in order to make it edible.

The problem is that nitrogen gas is to biological nitrogen as the dilute nectar of a flower is to honey. It requires processing to become most useful to you. You could die from a shortage of nitrogen atoms in your diet while carrying a full load of nitrogen gas in your lungs. Powerful tools are needed to convert that gas into more accessible forms, and only a handful of species possess them. These lucky few hold the keys to an inexhaustible trove of atoms that you and every other creature on Earth can't live without.

The feature that makes this element so hard to get also contributes to the color of the sky. Nitrogen seems to prefer the company of its own kind, at least when you come across it in its most common form. Other atmospheric gas atoms can also pair up in similar fashion by sharing one or more of their electrons in covalent bonds, but not nearly so tightly. The twin atoms in a hydrogen molecule are joined by one bond, and those in an oxygen molecule by two. But in dinitrogen the paired atoms clamp together with such a tenacious threefold grip that it becomes extremely difficult to separate them once they get hold of one another. If close friends can be said to be "joined at the hip," then the atoms in a nitrogen molecule are more like passionate lovers who are locked together from head to toe.

The electron clouds that surround and bind the atoms in air molecules such as dinitrogen are not perfectly rigid. If you could give such a molecule a vigorous shake the shell of electrons surrounding it would jostle slightly, and this responsiveness makes a remarkable thing happen when the sun rises on a clear morning. As the bees in the meadows of the world awaken and begin the day's harvest, the air molecules above them begin to hum as well, though not in waves of sound. Instead, the air buzzes with light.

The daytime color of Earth's nitrogenous sky is more than a mere backdrop to scenery: It also affects you physically. Recent investigations into the effects of light on human physiology show that exposure of your retinas to blue light can suppress your production of melatonin, a hormone that induces sleep. In days past this response to sky-filtered sunlight probably helped to keep your early ancestors alert and active between dawn and dusk through an unconscious, reflexive sensitivity that sleep researchers call "sightless vision." Today, however, some experts believe that the blue-enriched glows of computer screens, televisions, and other forms of artificial illumination that now shine around the clock may contribute to a wide spectrum of ailments from depression and insomnia to fatigue-related accidents.

To illustrate how the blueness of a sunlit sky is made, imagine the molecules in the air around you as a swarm of invisibly small honeybees—friendly honeybees, of course. Among them you'll find three kinds of movement going on at the same time. Winds blow the whole swarm here and there, and the individual bees also dart back and forth in a chaotic thermal dance. But much of the color of the sky emerges from the air molecules themselves, as the humming of bees emerges from their bodies.

The buzz of a bee comes from the rapid-fire shivering of tiny muscles which can power the beating of wings, shake pollen loose from a blossom, or warm the insect on a chilly morning. If you allow a good-natured bee to rest on your hand you can often feel the tingle of those vibrations. The atmosphere radiates light as bees radiate sound, but it does so with the help of electrons rather than muscles. When sunlight strikes dinitrogen, the molecule's electron cloud twitches rhythmically in response, and physicists who study that silent optical buzz sometimes refer to air molecules in essentially musical terms as "harmonic oscillators." One expert recently described the phenomenon to me, along with a strict warning not to bungle the explanation.

"Most people get this wrong," Craig Bohren told me, "even many scientists. As long as you don't make the mistake of saying that the sky is only blue or even predominantly blue, you are on the side of the angels."

With that surprising admonition, this well respected atmospheric scientist, now retired from Pennsylvania State University, led me through the basics of sky color.

"The nitrogen and oxygen molecules that dominate the atmosphere are much smaller than light waves, and this allows them to scatter sunlight in every direction when it hits them. They actually scatter all kinds of colors, but most notably the ones with shorter wavelengths." That would be blue, but also violet. So—why isn't the sky closer to violet?

"Well," he said with a chuckle, "in a sense it actually is, but you don't see it that way. Your eyes aren't as sensitive to violet as they are to blue, and your brain doesn't analyze light like a spectrophotometer would. What you see above you is partly of your own making, and you just give the name 'blue' to what you perceive up there."

According to Bohren's measurements, only about one-fifth of the light in a bright noontime sky is actually blue. "Even Einstein overlooked this detail when he investigated optical scattering, because he never looked at a sky-light spectrum."

And how, exactly, does an air molecule scatter light?

"If you're sure you want to get into it," he began after a pause, "it has to do with how the electron cloud responds when a rapidly oscillating light wave strikes it. The motion of the electrons relative to the nucleus generates an electromagnetic disturbance whose wavelength or color is related to the wavelength of the incoming light. An air molecule does this with many light waves at once, but its scattering of the blue and violet light is greater than that of, say, red or yellow."

In other words, molecular nitrogen can scatter the full palette of colors that stream out of the sun, much as a cell phone can reproduce many of the sounds of a symphony that a friend might send to you surreptitiously from a concert hall. But like the small speaker in your phone, which emits the higher pitches of a violin section more effectively than the deep booming of a bass drum, the tiny molecules of the atmosphere broadcast the relatively shorter waves of blue and violet light with greater fidelity. We are a bit tone-deaf when it comes to the high violets, but we can still get plenty of pleasure out of those equally glorious blues.

Multiply this process trillions of times per second throughout the lower atmosphere, add the limitations of human vision, and you end up perceiving something akin to a swarm of tiny, luminescent blue bees. The effect is so powerful that it can block your view of the stars during daytime, and you can read by scattered sky-illumination alone for quite some time after sundown. What appears to be an opaque dome overhead is more than just a passive ceiling: it glows with its own atomic light. But in the end, it is up to you to make it blue.

Although nitrogen atoms comprise only 3 percent of your total body mass, they are key components of molecules that make you look and act as you do. The carbohydrates and lipids in your body consist mostly of the three elements that comprise carbon dioxide and water, but in order to make the thousands of kinds of protein that keep you alive, you must also add nitrogen to the mix. Nitrogen atoms represent 10 to 15 percent of the dry mass of your muscles, and four nitrogen atoms cradle each rust-colored iron atom in the hemoglobin of your blood. All your enzymes, antibodies, and genes contain them, as do the ion pumps on your neurons and the cartilage in your nose. Without nitrogen in your diet, the most substantial parts of you that could be built from carbon, hydrogen, and oxygen alone would be your body fats and fluids.

The electron clouds that make nitrogen gas respond to sunlight also require a cell to have the molecular equivalent of a wrench in order to pry it apart and make other things from the atomic pieces. Luckily for us, some organisms do possess such tools. Microbes embedded in the roots of alder shrubs perform some of that work, but chief among the world's nitrogen fixers are colorful cyanobacteria that live in plankton or embedded in the crinkly tissues of lichens, and soil bacteria that colonize the roots of legumes such as alfalfa, clover, and beans. Because of these alliances among bacteria and their hosts, planting a field with alfalfa is like spraying it with fertilizer.

The molecular wrenches of nitrogen-fixing bacteria are iron-bearing enzymes called nitrogenases. A nitrogenase can snap dinitrogen in half, then decorate each atom with three hydrogens, thereby forming biologically useful ammonia. Unlike typical chemicals that are consumed or neutralized in reactions, enzymes such as nitrogenases are sturdy implements that perform their tasks over and over for as long as energy and raw materials are available.

Two potential sources of your nitrogen atoms. Left: *Mimosa pudica*, a tropical legume that houses nitrogen-fixing bacteria in its root nodules. Right: Photomicrograph of *Anabaena*, a lake-dwelling cyanobacterium that fixes nitrogen in enlarged cells such as the one in the center of this strand. *Photos by Curt Stager*

Through the ages nitrogen-fixing bacteria such as these have monopolized the production of nitrogen compounds that the rest of life on Earth has craved, stolen, and killed for. If they had been human traders of such a commodity, they would have been a fabulously wealthy cartel. Instead, root nodule bacteria have limited their demands to free accommodation in cozy, subterranean lodgings, and in the case of aquatic cyanobacteria, to chemical defenses that can punish creatures that try to eat them.

Apart from bacteria, the only other noteworthy nonhuman source of useful nitrogen is lightning. A bolt of lightning may be no thicker than your thumb, but it is hotter than the surface of the sun. The intense heat tears nitrogen molecules apart and allows oxygen to cling to the separated atoms. Each air-shattering stroke leaves a vapor trail of nitrogen oxide behind, which disperses into the atmosphere and eventually falls to the ground in rain or snow where plants can get at it and feed it into various food chains. You almost certainly carry atoms from the smoke of thunderbolts inside you right now, as do most other living things.

The invisibility of most of the nitrogen that you encounter in daily life makes it difficult to develop an inituitive awareness of your connections to it. But with a little scientific information and a healthy imagination, you can still probe beneath the surface of familiar scenes and situations for deeper atomic truths. For illustrative purposes, here is one such view of the nitrogen hidden in a setting that may be familiar to you—the Kansas tornado scene in the classic film *The Wizard of Oz*.

As the wind picks up under a darkening sky, a ropy tentacle of cloud snakes toward a sad-looking field, and between the field and funnel stand a house and barn. As the twister approaches, the people, chickens, and horses run for cover. Press the Pause button here. What are the nitrogen atoms doing?

This portion of the movie was filmed in black and white, so you couldn't see a blue dinitrogen sky above the long, tapered sock that makes such a convincing tornado on screen, even if it weren't cloudy. No matter—there is plenty to consider in the motion of the air rather than its color. Most of the force of those terrible winds represents the impacts of flying nitrogen molecules against the ground and the characters as they seek shelter. The fierce flow was largely driven by the differential thermal movements of nitrogen molecules farther upwind, as the heat of the sun sped their dances in rising, swelling parcels of air. The collision of warmer, faster-dancing air masses with cooler, slower-dancing ones over Kansas spawned this devastating storm.

Flashes of lightning could be edited into those cinematic clouds as well. Every explosive bolt would leave a streak of nitrogen oxide in the turbulent air. Some of the oxides would dissolve into raindrops and soak into the soil, later to help sustain whatever crops survived the storm.

Unseen in this view are the nitrogen-rich vapors rising from the field. Bacteria in the soil have been eating leftovers from last year's harvests and manure piles and unwinding their proteins into the gases from which they originated. Bacteria beneath an outhouse that stands behind the barn are doing the same.

Nitrogen atoms are hidden in the dry stalks in the field as well, lingering in the remains of once-green chlorophyll. They are even more abundant in the keratin feathers of the chickens as they run squawking through the barnyard.

The film resumes with Auntie Em stepping out into the gale to call for Dorothy as the dinitrogen wind whips the nitrogen-bearing keratin of her gray hair. Dorothy herself, stumbling in the blinding gusts, is too busy coordinating her nitrogen-rich muscles to hear the nitrogen-borne sound waves of her aunt's cries. Her little dog, Toto, is faring a bit better, however, because his sharp keratin claws give him a better grip on the ground.

Similar nitrogenous connections operate all around you in the modern world, but some new ones have also entered the picture now. The spark plugs in the engines of our motor vehicles resemble miniature thunderstorms, and they fix nitrogen just as lightning does. You inhale close to three thousand gallons of nitrogen-laden air per day, on average, a volume equivalent to that of three dozen large bathtubs. A typical compact car, by comparison, inhales thirty times as much in an hour, and if you drive it 12,500 miles over the course of a year it can fix roughly eighteen pounds of nitrogen oxides. Multiply that respiratory rate by four for a large truck and by closer to a hundred for an airliner, and you can imagine why vehicle exhaust, along with the fiery innards of coal-fired power plants, has become a major component of the global nitrogen cycle.

The Canadian meteorologist Lewis Poulin estimated that traffic in and around Montreal in 2001 processed about 175 times more air per day than the city's 1.8 million inhabitants did. But unlike the nitrogen molecules that people exhale unaltered, many of the ones that pass through those superhot engine cylinders emerge as nitrogen oxides. Scientists with the United States Environmental Protection Agency have shown that every major city on the Atlantic Coast sends a smoggy plume of nitrogen-enriched gases far out to sea on the prevailing westerly winds, along with other waste fumes. And the eastern cities, in turn, also lie downwind of similar plumes arising from cities and highways farther inland.

Roughly half of the planet's biological nitrogen, including most of the nitrogen in your own body, is now snatched from the air by fossil fuel combustion and, above all, by a process that was first harnessed on an industrial scale to supply Germany with explosives during World War I. This artificially fixed nitrogen can now end up in anything from TNT (tri*nitro*toluene) to laughing gas (*nitrous* oxide). Even fertilizer (ammonium *nitrate*) has been used in the full spectrum of human endeavors from famine prevention to the Oklahoma City bombing. All these things are the legacy of one man, placing him among the most influential figures in human history.

Fritz Haber was born in 1868 in Breslau, East Prussia (now Wrocław, Poland), the son of a merchant. According to biographers, he was a brilliant scientist, a lousy husband, and an ambitious patriot. This combination of traits helps to explain much of the trajectory of his life. As he himself described his philosophy of hard work and civic duty, "I don't want to rust out, I'd rather *wear* out."

Like Einstein, Haber was a German Jew. But unlike Einstein, he converted to Christianity in order to further his advancement in an increasingly anti-Semitic society. Around 1905 he found a way to make liquid ammonia by combining nitrogen and hydrogen gas under high temperature and pressure, using an iron catalyst to push the reactions along in much the same manner that bacterial nitrogenase does. By 1909 he and the British physical chemist Robert Le Rossignol had developed a high-pressure apparatus that dripped ammonia in potentially useful quantities. Although he was already well on his way to prominence in the scientific community, perfecting this process made Haber into a public celebrity.

At that time Germany's largest sources of nitrogen for fertilizer and explosives were layered crusts of sodium nitrate found in desert salt flats in northern Chile. Such deposits are extremely rare, and their origins remain somewhat mysterious. Sodium nitrate dissolves easily in water, so it could only accumulate to such an extent in hyperarid environments such as the Chilean desert. In a report published by the U.S. Department of the Interior in 1981, the geologist George Erickson concluded that the strange deposits had accumulated from sea spray, volcanic emissions, and the weathering of rocks and soils over millions of years. Almost any other location would have been too wet and vegetated to allow so much soluble, biologically useful nitrogen to build up in this manner. To Erickson, finding those foot-thick layers of fertilizer, some of them exceeding 50 percent purity, must have been almost as surprising as finding a huge plain of table sugar lying out in the open without it washing away or attracting hungry ants.

Whatever its origin, Chilean saltpeter (from the Latin words *sal* and *petra*, referring to salt and stone) enriched those who exploited it since the early 1800s, and by the early 1900s the United States and Britain had come to depend upon the great saltpeter flats as well. As the shadow of World War I loomed darker, however, the vulnerability of that single source and the long voyage from Chile turned Haber's invention into a game changer for Germany.

When his fellow chemist Carl Bosch scaled the reactions up to industrial levels between 1909 and 1913, what then became known as the Haber-Bosch process turned the atmosphere itself into a gigantic nitrogen mine. Those who focused on its agricultural uses spoke in glowing terms of making "bread from the air." But ammonia is also readily converted to nitric acid for the manufacture of explosives, and in the midst of war the new procedure couldn't have been more desirable to the homeland. On the other hand, it also allowed the bloody conflict to continue longer than it otherwise would have if nitrogen resources had remained limited to geological deposits and root nodules.

How can a substance that promotes life when used as fertilizer also destroy it so violently? Nitrate molecules feed their own oxygen atoms to a fire so rapidly that a normally slow burn becomes a savage blast. Mix potassium nitrate with charcoal and sulfur in varying proportions, for example, and you get various forms of gunpowder that have long been used in weapons and fireworks.

When the eighteenth-century French chemist Jean-Antoine Chaptal gave nitrogen its elemental name (*nitrogène*, "source of nitre"), he was referring to its presence in saltpeter, which was then commonly called "nitre." Each nitrate subunit in a molecule of potassium nitrate consists of a nitrogen atom joined to three oxygen atoms, and when a spark strikes it the nitre in gunpowder unloads its oxygen directly onto the flammable compounds around it. Then everything seems to happen at once. The flammables erupt into carbon dioxide, water vapor, and sulfurous gases, all of which expand at terrible speed along with heat and light. As the powder explosion in a gun pushes a bullet out of its barrel, nitrogen atoms that lost their oxygen baggage also fly out as nitrogen gas, where they blend back into the air from which they were originally fixed.

When you digest the proteins in your food, you eventually excrete their nitrogenous remains in the form of urea. Bacteria may later convert that urea into ammonia, which in turn, can become oxidized into nitrate suitable for explosives. During the American Civil War, the resource-starved Confederate army was sometimes reduced to making gunpowder from wood ash (for the potassium) and urea (for the nitrate). According to the Web site of the musician and author Rickey Pittman, every source of urea-based nitrate was exploited, from barns to chamber pots.

Pittman quoted a public notice posted by the government agent John Haralson as saying "The ladies of Selma are respectfully requested to preserve the chamber lye to be collected for the purpose of making nitre. A barrel will be sent around daily to collect it." Poets quickly seized upon the tale, and one song of the day ended with this bawdy verse:

> *John Haralson, John Haralson, pray do invent a neater,*
> *And somewhat less immodest way of making your saltpeter.*
> *For 'tis an awful idea, John, gunpowdery and cranky,*
> *That when a lady lifts her skirts she's killing off a Yankee.*

By fixing ammonia directly from the air, Fritz Haber revolutionized modern warfare and became a national hero. You might expect his celebrity status and contributions to the fatherland to have played well at home with his wife, Clara, who was herself a talented chemist. But their relationship was strained by Haber's seeming eagerness to fulfill any demands that his country might make of him, regardless of their ethical implications.

Haber did not share Einstein's pacifist philosophy and willingly applied his research to military ends. In his view, "A scientist belongs to his country in times of war and to all mankind in times of peace." As the director of the Kaiser Wilhelm Institute for Physical and Electrochemistry, which is now named after him, Haber helped to develop and supervise the use of poison gas as a way to drive Allied troops from their trenches. He often said that he hoped such devastating tactics would bring a speedy victory to Germany and would therefore ultimately save lives that might otherwise be lost in a more prolonged conflict. He was unfortunately mistaken in that regard.

For Clara, her husband's activities had begun to cross the line into complicity with evil. According to an article in *Smithsonian* magazine by the author Gilbert King, she protested his work in public as well as in private, calling it "barbarity" and "a perversion of the ideals of science." He responded by accusing her of treason, further weakening their already-troubled marriage.

The last straw, it appears, came in the spring of 1915 after Haber personally supervised the first use of chlorine gas on Allied troops in Flanders. One Canadian soldier who survived later described being

Left: Clara Immerwahr Haber. Right: Fritz Haber. *Courtesy of Archiv der Max-Planck-Gesellschaft, Berlin-Dahlem*

gassed by German chlorine as "an equivalent death to drowning only on dry land. The effects are . . . a knife edge of pain in the lungs and the coughing up of a greenish froth . . . ending finally in insensibility and death." Shortly after hearing of the gas attack and the ghoulish suffering that resulted, Clara walked into the family courtyard, aimed a pistol at her chest, and pulled the trigger. Their thirteen-year-old son, Hermann, who found her as she lay dying, was left to grieve alone when Haber departed the next morning in order to oversee gas releases on the Eastern Front.

It is tempting to blame Haber for his wife's suicide, though Clara left no explanation for her fatal decision. Certainly his family life left much to be desired, at least in part because of his willingness to put ambition and country above all else. But when the war ended, Haber redirected his efforts to support more peaceful goals. He spent much of the 1920s in an attempt to help Germany pay war reparations by extracting gold from seawater, and he proposed using fertilizers and pesticides to turn the Sudanese desert into an agricultural Eden.

In the end, however, the nation to which Haber had devoted himself betrayed him. After Hitler came to power in 1933, Nazi race laws began to purge Jewish scientists from even the most prestigious posts and drove Einstein and others to emigrate to America. One day the former hero of nitrogen fixation went to work at the institute only to be turned away by a porter who announced, "The Jew Haber is not allowed in here." He resigned and exiled himself to England, where—not surprisingly—he was shunned by British scientists for his involvement in gas warfare, and then moved to Switzerland. According to the biographer Morris Goran, a friend described Haber in his postwar years as "seventy-five percent dead." Fritz Haber succumbed to heart disease in 1934, a broken man. After his death his old friend Einstein reportedly concluded, "Haber's life was the tragedy of the German Jew—the tragedy of unrequited love."

Some say that it was fortunate for him that he did not live to see what arose from his work after his passing. By the cruelest of ironies, research by Haber and his associates at the Kaiser Wilhelm Institute led to the invention of Zyklon B, a nitrogen-bearing pesticide that was originally intended to protect crops but that would later be used in Nazi gas chambers to kill millions of Jews, including some of Haber's former colleagues and extended family. And in 1946 Hermann Haber died by suicide, reportedly because of his shame over his father's wartime research.

How does one weigh the life of such a person?

In an address to Parliament in 1918, Winston Churchill remarked:

> It is a very strange thing to reflect that but for the invention of Professor Haber the Germans could not have continued the War after their original stack of nitrates was exhausted. The invention of this single man has enabled them . . . not only to maintain an almost unlimited supply of explosives for all purposes, but to provide amply for the needs of agriculture in chemical manures. It is a remarkable fact, and shows on what obscure and accidental incidents the fortunes possibly of the whole world may turn in these days of scientific discovery.

If you go strictly by the numbers, perhaps Haber's effects on recent history yielded a net positive. More than one hundred million war dead and great devastation throughout the last century might in theory be counterbalanced by the effects of air-derived fertilizer on human hunger. It has been estimated that as many as half of today's seven billion people couldn't even exist if it weren't for artificially fixed nitrogen because there simply would not be enough usable nitrogen atoms available to build and maintain their bodies with. The outcomes of such equations depend on how the factors are weighted, and it could also be argued that more people can mean more pollution and strife in the world.

However you judge it, the Haber-Bosch process certainly liberated us from the constraints of the ancient bacterial cartel. Global commercial ammonia production currently exceeds one hundred million tons per year, and like other nitrogen fixers we also pass our nitrogenous wastes out into our surroundings in many forms. Thanks to the science and technology of a modern civilization that Haber helped to create, we can now explore those atomic connections in unprecedented and fascinating detail.

It is often said that "you are what you eat." Some pioneering ecologists are now demonstrating this principle in the wilds of North America's Pacific Northwest. Their primary subjects are not people, however, but salmon.

The annual migration of sockeye salmon upriver from the sea to spawn and die in remote lakes is a spectacular example of determination in the face of adversity. It has also been much more than that to residents of forested watersheds along that rugged coast for thousands of years. Salmon, quite literally, are part of the fabric of life there. And nitrogen, or more specifically a heavy stable isotope of it, is allowing scientists to tell new versions of that ancient story.

When you see a river seething with salmon, it is easy to lose track of individuals amid the chaos. Because of this problem, biologists use marker tags to track single fish in order to work out where they and their companions go, how long they live, and how many are caught relative to the population as a whole. But if you want to dig deeper into the story of salmon, normal labels won't suffice. When a fish dies, its tag falls off and is lost. A new kind of marker now enables us to follow a salmon's nitrogen atoms far beyond the fish to the occupants of surrounding forests as well as the forests themselves.

According to one biologist at the University of Victoria, British Columbia, who studies the atomic nature of salmon for a living, it all starts with the smell of rotting seaweed.

"You know that odor that hits you when you walk down to the beach," Tom Reimchen said when I spoke to him by phone. "When seaweed decomposes it releases ammonia and other gases as bacteria break it down. This sort of thing also happens in the rest of the ocean whenever something decays." For Reimchen and his colleagues, the isotopic composition of those molecules permits the tracking of salmon atoms through ecosystems.

"There are two main kinds of nitrogen atom out there," he continued. "One of them is a little heavier than the other, and it escapes into the air a little less easily. This difference makes the heavier isotope build up in the oceans."

Although you might think that atoms are as uniform as a school of fish seems to be, the existence of isotopes shows otherwise. Nitrogen-15 carries one more neutron than N-14, and being a little overweight, N-15 is a slightly more reluctant traveler than its lighter cousin. To people like Reimchen that reluctance is a good thing.

"Heavy N-15 vibrates less than N-14, so when it binds to other atoms the bonds tend to be more stable and harder to break. The ammonia rising out of rotten seaweed is enriched in N-14 because microbes have an easier time breaking it loose from proteins, which leaves more N-15 behind." This also happens in marine food chains, with every step from prey to predator increasing the N-15 load of the organism. Single-celled algae consume the nitrogen compounds of dead seaweed, and the process of converting those food molecules into algal cells concentrates the N-15 relative to N-14. Tiny planktonic creatures eat the algae and concentrate the N-15 some more, small fish eat them, and so on. Sockeye salmon exist near the top of their food chain, and they feed only in the N-15 enriched oceans, so they carry distinctively large fractions of N-15 in their bodies. Scientists routinely find N-15 concentrations more than ten times higher than atmospheric or terrestrial values within the tissues of sockeyes.

"A fish that feeds in the ocean contains more N-15 than a freshwater fish does," Reimchen explained, "and the same is true of anything else that eats food from the sea rather than from the land or a typical river or lake." As a result, high concentrations of N-15 in body tissues can reveal subtle atomic ties to the ocean in a previously undocumented ecosystem, the "salmon forest."

Shape-shifting has long been a mainstay of mythology throughout the world. In such traditions witches become black cats, hairy men become werewolves, and vampires become bats. We may chuckle at those tales, but scientists such as Reimchen don't blink an eye. In the realm of atoms, fish become bears, wolves, birds, and human beings, and deep oceans can become lush forests.

The Queen Charlotte Islands (more formally known as Haida Gwaii) lie just off the coast of British Columbia, and they attract migratory salmon into their rivers year after year. In an article published in *Ecoforestry*, Reimchen reported that a single Queen Charlotte black bear can capture as many as seven hundred salmon during the six-week spawning period in late summer and early autumn. Normally a bear carries its prize off into the woods to dine in private, and it tends to eat mostly the choicest bits, leaving the rest of the carcass on the ground while it returns to the hunt. In the realm of atoms, food rarely goes to waste, and the leftovers of bears' feasts soon become dinner for eagles, ravens, and gulls. Days later any uneaten soft parts and animal droppings are bonanzas for beetles, flies, and bacteria that turn them into nitrogenous wastes. The decaying refuse in the riverside forest produces, as Reimchen put it, "a highly odiferous riparian zone."

As with black bears, a grizzly's nitrogen intake can also come almost exclusively from fish protein during the annual salmon migration in British Columbia. After a lunge, a clamping of jaws, and a shaking of heavy wet fur, a successful hunter drags its prey into the bushes. Tearing open the skin to expose orange-pink meat, the bear gulps a mouthful of fish protein into its stomach, where enzymes chop it into smaller, nitrogen-bearing amino acids. Amino acids from the meal ride the predator's bloodstream to the liver, where most of them are processed into the raw materials for muscles, sinews, and other body parts. Anything above the daily requirement for growth, tissue replacement, and such things is broken down into ammonia, which is then converted into urea for disposal. Nine out of ten of those discarded nitrogen compounds are filtered out by the kidneys and excreted in urine, and most of the rest exit through solid waste. Similar things would also happen inside you if you should likewise take a bite of savory salmon and make those fish atoms your own.

If the bear is a lactating mother, then much of the unused nitrogen goes into milk protein instead, but the body normally uses most of the amino acids for producing fur and other items of bear anatomy. In one study 80 percent of the nitrogen in bear fur could be traced back to the ocean. This atomic signal was present in newly emerging hair shafts not only during the fall salmon migration, but also in spring and early summer, when the bears were eating more vegetation than fish. Oceanic nitrogen, it turns out, suffuses the entire forest, not just the bears.

Plants use nitrogen atoms in the light-trapping chlorophyll that turns them green, as well as in enzymes, cell walls, and fragrances. Near salmon streams much of the nitrogen that plants absorb from the soil comes from bear-processed fish proteins, and Reimchen and his colleagues have measured high concentrations of N-15 in local huckleberries, wild azaleas, and the aptly named salmonberries, which reveal in atomic script the isotopic signature of the sea. Similar studies have found the same pattern in watersheds all along the northern Pacific Coast, in which the N-15 content of foliage reflects the proximity of plants to a stream, the density of fish, and the abundance of bears.

Unfortunately, less desirable gifts from the sea also travel through these food webs with the nitrogen atoms. According to a study by the environmental scientist Jennie Christensen and her colleagues, air-

borne pollutants that fall into the Pacific Ocean can eventually con-
taminate salmon, as well. They concluded that migratory salmon
deliver up to 70 percent of the organochlorine pesticides and 90 percent of the toxic PCBs found in fish-eating grizzlies, "thereby inextri-
cably linking these terrestrial predators to contaminants from the
North Pacific Ocean."

Meanwhile, back at the carcass, other bits of salmon nitrogen are
trotting away in the stomachs of wolves, weasels, and foxes, or flying
up into the trees with the eagles and ravens. Some make detours
through carrion-feeding insects to flycatchers and warblers before
falling back to the ground in bird droppings. Even the nitrogen atoms
in the leftover fish skeletons slowly escape as bacteria release ammonia
and nitrate into the soil, where fungal fibers distribute them among
the roots of the forest.

The trunks of the spruce and hemlocks lining Reimchen's study
streams contain more N-15 than those of trees growing farther upslope
or upstream of waterfalls that block salmon migrations. As much
as 40 percent of the nitrogen in riverside vegetation at the British
Columbia sites may come from the bodies of fish by way of bears, but
at one coastal stream where the highest salmon densities were found,
Reimchen estimated that some of the heaviest migration years see
more than three-quarters of the annual nitrogen budgets of the local
spruce trees supplied by fish.

Studies show that salmon nitrogen can make plants grow faster
or more profusely, and some investigators have used tree rings as re-
corders of salmon abundance through time. Most fisheries studies
cover only a few years or decades, but spruce, fir, and hemlock trees
can live for centuries, laying down annual bands of new wood in the
outer layers of their trunks. By measuring the abundance of N-15
ring by ring in streamside trees, researchers from the University of
Washington found that salmon populations fluctuated a great deal
during the last 350 years, even before modern human impacts began.
But the longest records of salmon history come from lakes in which
survivors of the predator gauntlet end their epic journeys.

As sockeyes work their way up into the headwaters of their home streams, the males lose their silvery sheen and develop red flanks, green heads, and wickedly curved jaws in preparation for courtship and turf battles. When the fish reach a lake, they begin to spawn. Females lay eggs in the shallows, and males fight for a chance to drop their milky gametes on the nests before dying. Soon the lake floor is littered with battered carcasses, which decompose and release nitrogen atoms into an aquatic version of the salmon forests.

Algae consume the nitrogen compounds and then become food for microscopic zooplankton, producing bite-size meals for newly hatched salmon. The youngsters spend several years fattening up on the recycled atoms of their forebears before making their own round trips to and from the Pacific. And all the while, thin films of detritus and dead plankton accumulate in the soft mud beneath the lake like annual bands in a tree trunk. These layered sediment archives can represent thousands of years of salmon history.

In a paper published in *Nature* in 2002, ecologist Bruce Finney and his colleagues described what they found in sediment cores from Karluk Lake, a wild sockeye nursery on Kodiak Island, Alaska. During the last two millennia, sedimentary N-15 values rose and fell along with the glassy shells of diatom algae that tend to grow most prolifically in years when salmon migrations are largest. But more surprising than the length and detail of the records was what they revealed about today's populations.

Long-term declines such as those that are now under way are not unique to our times, and an even larger crash that occurred two thousand years ago lasted for several centuries. We don't yet know what caused it, although Finney's team suspected that it had something to do with fluctuations in climates, Pacific currents, and sea surface temperatures.

Many other questions also emerge from these discoveries. How abundant are salmon "supposed" to be? Were old reports of rivers so full that you could cross on the backs of the fish just "big fish" tales? When commercial fishing first started in the region, salmon were more abundant than they had been for thousands of years. Did that coincidence leave us with a false impression of what to expect in the future?

Such investigations show that we still have a lot to learn about salmon and their place in the nitrogen economy of the Pacific Northwest. And not surprisingly, similar studies are beginning to uncover a great deal about our own nitrogenous connections to the world as well.

Through most of human history, the atomic exchanges between our ancestors and their surroundings were much like those in a salmon forest, and they can still be traced through human bodies with isotopic markers just as they are in Reimchen's research.

Archaeologists are using N-15 analyses of collagen, the most abundant protein in bone, to reconstruct the diets of people who died long ago. In one such study, the archaeologist Erle Nelson and his colleagues exhumed the skeletons of early Norse settlers in southern Greenland in order to help determine why their settlements failed during the fifteenth century. Some experts have proposed that a natural cooling event, the so-called Little Ice Age, made conditions too harsh. Others have surmised that a stubborn refusal to adopt the hunting and fishing habits of the local Inuit led to death by soil depletion and crop failures. In 2012 Nelson's team probed the atomic contents of Greenlandic graveyards in order to test these competing ideas.

Their isotopic analyses showed that the Greenlandic Norse ate quite a bit of fish and seal meat, thereby weakening claims that climate or crop failures did them in. As with the bears of the Pacific Northwest, people who eat marine creatures carry more N-15 within them than do people whose nitrogen atoms come from crops and domestic livestock. In the Greenland study all the skeletons contained more N-15 than they would have if the settlers had lived, as it were, on bread alone.

All the skeletons, that is, except one. According to local documents, this individual was born in Scandinavia and had joined the colonists only a short time before he died. As a result his bones still carried many of the agricultural nitrogen atoms from meals that he consumed during his earlier years back in Norway. The relatively low N-15 contents of his remains confirmed that he had not yet replaced all of the nitrogen in his body with local food atoms at the time of his death.

According to the authors, the nitrogen isotopes of those early settlers showed that the demise of the Greenland colonies was probably not so much a sudden collapse as a slow drain-off. In a press release from the University of Copenhagen, the anthropologist Niels Lynnerup explained: "Nothing suggests that the Norse disappeared as a result of a natural disaster. If anything, they might have become bored with eating seals out on the edge of the world." Longing for the richer cultural and social environments of Scandinavia may simply have lured so many young people away that the remaining villagers finally gave up and left, too.

Digesting the proteins of another organism tends to leave more of the heavy N-15 atoms behind in one's tissues, and this typically makes the N-15 concentration of a predator's body higher than that of its prey. The same principle applies to us. According to a paper published in *Organic Geochemistry* in 1997, a wonderful manifestation of this kind of atomic linkage can be found in the nutritional connections between nursing mothers and their infants.

When you were still a fetus, all your atoms came from your mother's body. No eating, drinking, or breathing for you yet, only the cyclic pumping of fluids between placenta and umbilical cord. You were essentially a piece of your mom then, and although you breathed your own oxygen after you were born, you would still have gotten your other atoms through mother's milk during infancy unless you were raised on bottled formula.

But although your first atoms came from your mother, the body that you built from those atoms was slightly enriched with N-15 in much the same manner that a bear's proteins are enriched relative to the salmon it eats. In other words, you were not only a part of your mother: You were also feeding on her. Call this relationship predation, parasitism, or even cannibalism if you like; it was all the same to your nitrogen isotopes, which recorded what you were doing and wrote it into the molecular archives of your proteins.

When investigators gently trimmed the fingernails of breast-fed newborns and compared them to fingernail clippings from their mothers, they found the telltale isotopic signs of that maternal sacrifice. A captured salmon unwillingly feeds its N-15 to a hungry bear, but nursing mothers freely offer their own atoms to their babies. In keeping with the age-old rule of food chains, the proteins of the infants in the study were a little richer in N-15 than those of their moms, and they remained so until the children were weaned and their isotopic contents shifted into closer alignment with the atoms of the larger world that would henceforth sustain them. The first isotopic signs of independence appeared in clippings after two to three months, the time it takes for a newborn's fingernail to grow from cuticle to fingertip. Similar isotopic studies show that the course of a pregnancy can also be traced along the lengths of a mother's hair filaments as nitrogen atoms drain from her body into her child.

In many ways the intimate connections between mother and child are also mirrored in our atomic ties to the Earth. Even after we pull away from our early umbilical linkage, we still live embedded in the same global reservoir of recycled atoms. But today those atomic connections join us not only to living things but also to machines.

At the time of this writing, our vehicles, farms, and industries produce about half of the biologically useful nitrogen on Earth. One study published in *Science* in 2010 reported that the Haber-Bosch process alone now matches the entire nitrogen-fixation output of the oceans and exceeds the microbial fixation on land. We live in a world that is fundamentally different from the ones that our distant ancestors knew. No, most of us are not in Kansas anymore.

The people in the black-and-white world of *The Wizard of Oz* lived off whatever their farm could provide, and the nitrogen in their bodies came from their livestock, crops, and the occasional lightning bolt. From soil bacteria to plants and farm residents and back again, nitrogen atoms returned over and over to the same soil as horse teams plowed it and the hired help spread hog droppings and other wastes on it. Not a perfectly idyllic existence by any means in those hard times, even without a twister about to strike, but one from which we still have important lessons to relearn in greater depth. Those people were part of an elemental cycle that is now largely broken in present-day Kansas as well as in much of the rest of the world.

If you are like most Americans, you obtain nearly all of your food from supermarkets that are supplied by industrial-scale farms. It would therefore be safe to assume that most of the nitrogen atoms in your own body were artificially fixed a long distance from where you live. Most of the energy for that fixation came from massive amounts of nonrenewable fossil fuel, as did the energy for the transportation of the food to your local stores and then to your table. Many of those costly nitrogen compounds were never absorbed by crops in the fields upon which they were sprayed, and others escaped from livestock in wastes that ended up in storage lagoons or sewage treatment facilities where denitrifying bacteria returned them to the air as inert nitrogen gas. Such processes open enormous gaps in the human-centered portions of the global nitrogen cycle that must be filled by yet more energy-intensive fixation and transport.

One could argue that this is how it must be, that this is now the only way to keep so many people alive and free from the agony of hunger. But finding ways to more effectively close the loop on our nitrogen cycle is both feasible and beneficial in the long run. Returning animal wastes to the soil, replenishing fields with nitrogen-fixing crops, and more carefully adjusting the timing and amounts of fertilizers that we spread on our fields are among the many relatively simple suggestions now in circulation.

For hundreds of thousands of years our ancestors absorbed, used, and released these same atoms back into the same finite pool, trusting the world to provide them with what they needed and to dilute or dispose of their wastes. The Haber-Bosch process has propelled us into this present century as the dominant source of living nitrogen on Earth, and as our ability to draw nourishment from the atmosphere increases, so too does our influence on the global nitrogen cycle.

The ecologist David Schindler described our situation succinctly in an interview for the University of Washington. "The world, for nitrogen," he said, "is a much smaller place than we'd assumed." The nitrogen atoms that we use remain on the planet with us when we discard them, and they now do so in ever-growing quantities along with our other wastes. In addition to feeding our crops, our nitrogenous emissions also trigger unwanted algae blooms, acidify rain and snow, and help to toxify urban smog. They even contribute to climate change, because the nitrous oxide that is released from overfertilized fields, lawns, golf courses, and fossil fuel combustion is a greenhouse gas. As the ecologist Alex Wolfe noted in an online interview, "The global change debate is dominated by discussions of carbon emissions . . . [but] the global nitrogen cycle has been far more perturbed by humanity than that of carbon."

How we balance our need for fixed nitrogen against its effects on air and water quality, as well as our evolving ability to feed or to kill one another, will be one of the great unfolding stories of our journey forward as a rapidly maturing, sentient species. And however that story develops in the future, the ambiguous legacy of Fritz Haber will surely continue to play a role in it, too.

7

Bones and Stones

We abuse land because we see it as a commodity belonging to us. When we see land as a community to which we belong, we may begin to use it with love and respect.
—Aldo Leopold

By blending water and minerals from below with sunlight and CO_2 from above, green plants link the earth to the sky.
—Fritjof Capra

It's not every day that someone presents you with a bone from the hand of a prehuman ancestor.

When it happened to me in 1988, I was a young scientist on assignment for *National Geographic* magazine in northern Kenya. I had arrived at the National Museum in Nairobi the previous day, thrilled to meet the paleoanthropologist Richard Leakey, who had been featured in television documentaries and magazine articles that I had enjoyed since childhood. Soon afterward he would lose the lower portions of his legs in the crash of a small plane like the one that would take me to one of his remote study sites. But this flight ended safely on a dusty strip of sand and gravel on the parched western shore of Lake Turkana, where I would encounter signs of the atomic connections among all of humanity and the earth itself.

From the air Lake Turkana resembles a muddy greenish-blue ribbon of water stretching 170 miles from the mouth of the Omo River in the north to a barrier of lava rock and volcanic cinder cones in the south. The shape of the lake reflects that of the Great Rift Valley, a system of gigantic tectonic cracks in the African continent that extends from the Red Sea to Malawi. The inland sea is brackish, so it is of little use to humans as a source of drinking water. But for the hardy people who wander the harsh desert terrain with their cattle, goats, and camels, the namesake Turkana and other pastoralists, this is home. And for scientists such as Leakey and his colleagues, this place also opens windows on the full sweep of human history.

As we approached the makeshift landing strip at the West Turkana fossil camp, I could see that the lake used to be much larger. Parallel stripes from former beaches graded into the forbidding landscape of thornbushes, low bluffs, and dry gulches.

"The farther you walk from the shoreline," the pilot explained as he banked the plane over a small cluster of tents, "the farther back in time you go. These beds were laid down a few thousand years ago, but the ones over there are closer to two million years old. If you keep on going all the way out to those mountains on the horizon, you'll be walking on dinosaur fossils."

Lake sediments are good hunting grounds for old bones because they cover and shield them from gnawing animals and the erosive forces of weather. Over time, minerals from the sediments also replace some or all of the original material with sturdier stuff that won't decay even if the fossil is later unearthed. This alone would make West Turkana of interest to scientists, but a special few of the fossils that lie buried here are so unusual that they make headlines around the world when they are discovered. Here lie the petrified remains of some of our closest hominid relatives.

As the geologist Frank Brown led me from the tent camp to a new excavation site a short distance away, signs of the deep past were everywhere. He paused to describe some of them to me.

"See that sandstone there?" I nodded. "The ripples in it show that this spot used to lie under shallow water."

"And what do you think those little bowl-shaped depressions in it might be?" I hadn't a clue.

"Nests. That's where fish scooped the sand away to shelter their eggs, just like some of their descendants do in African lakes today. Now, what about these lumpy formations over here?"

I stepped gingerly among what appeared to be half-buried skulls, each one roughly the size of an overturned washtub.

"These are stromatolites, a kind of microbial reef. Back when the lake covered this spot, mats of algae and bacteria built them up layer by layer from minerals in the water."

Even more interesting to me were the flat, teardrop-shaped stones scattered so thickly that we couldn't avoid stepping on them. "Are these what I think they are?"

"Yes, hand axes. Not much use to an anthropologist now that they're no longer in their original layers and can't be properly dated. But judging from the look of them, these were probably made by *Homo erectus*. Speaking of which, here we are."

The site resembled a patio cut into the side of a bluff. There on a flat stretch of densely packed sediment, a middle-aged Kenyan man in shorts and a checkered shirt knelt on a scrap of foam padding. Holding a small tool that I took to be an ice pick, he seemed to be scratching at something on the ground in front of him. As we came closer he rose somewhat stiffly to greet us, and I could see what he had been doing. Next to the pad was a shallow patch that had been carefully pecked away and sieved for bone fragments. This, my guide informed me, was Kamoya Kimeu, a widely respected fossil expert.

"Welcome," he said, extending a dusty hand. "As you can see, this is still a work in progress." He turned to gesture at the bluff wall, and I caught my breath. Kimeu saw, smiled, and brandished his hand spike. "Much of this excavation was made inch by inch with tools like this one. It has been a lot of digging since I found the first fragments here four years ago, but the work has been worth it. Most of the team has left now, but I'm staying on until every last piece has been found."

With that he held out his other hand. Cradled in it was a slender brown finger bone. "*Homo erectus*," he said. "This belonged to one of my ancestors. We're guessing that he was only about ten to twelve years old when he died a million and a half years ago."

Then, after pausing long enough for that message to sink in, he added with a twinkle in his eye, "He was one of your own ancestors, too."

Kamoya Kimeu excavating the "Turkana Boy" in northern Kenya, 1988. *Photos by Curt Stager*

Pulling hominid bones out of the ground can tell you a lot about yourself. Perhaps most obviously it reminds you that you have a strikingly similar skeleton inside of you. It also shows that your lineage runs deep into geologic time, that your body has prehuman roots, and that death has always been an inevitable result of having lived. Sometimes it can make you wonder if someone in a far-off future age might excavate your own remains in this manner. And as structures comprised of minerals, old bones and teeth can also remind you that your skeleton is a highly processed version of dirt.

Through the ages teeth have served mammals in much the same manner that handcrafted tools have served humankind. Fangs stab and grip, molars slice and grind, incisors snip and nibble. But bones also do life-sustaining work. They shield your organs from impact. They provide the anchorage and leverage to muscles that enable you to stand, walk, speak, and manipulate things. And they also store and dispense calcium and phosphorus, two vital elements that comprise most of the mass of your skeleton and teeth in the form of a mineral that geologists call apatite.

When embedded in a pebble of granite, apatite can be a tiny blue gem, and in a marine deposit it can be a glossy black nugget. But in the shells of turtles, the antlers of deer, the tusks of elephants, and inside of you, most biological apatite is ivory-white. Each microscopic crystal resembles a stack of atomic beads in which calcium and phosphorus are joined with oxygen. Irregularities in the crystal lattice also make room for additional visitors from the atomic world, making apatite one of the most variable minerals known. Its name comes from a Greek word meaning "deceit," which refers to the flexibility of form that allows it to play multiple roles in your body.

Calcium and phosphorus can team up with many elements, with or without one another. The chalky carbonate cements in eggshells, seashells, and coral reefs represent alliances among calcium, carbon, and oxygen atoms, and lone calcium ions also help to coordinate the beating of your heart and the clotting of your blood. Phosphorus strengthens the backbones of your genes and aids in the storage and release of energy from your food. But the apatite partnership of calcium and phosphorus more strongly supports your body in the most literal sense.

The atoms of the bone in the Turkana boy's finger were not created by the boy himself, of course. Like your own atoms, they formed in stars before the birth of the solar system. Billions of years later, herbs and shrubs coaxed them from soils and passed them into food chains. Eventually they crystallized in the finger of this young man shortly before rejoining their geologic companions along the shores of Lake Turkana. Now we value that ancient bone for scientific purposes as the late owner valued it on a more personal level.

Such is the legacy of calcium and phosphorus, which link us to the erosion of mountains and the growth of forests. Ultimately they also link you and me to one another, as well as to Kamoya Kimeu and the unfortunate hominid boy whose skeleton rejoined the African soil whence it came.

The atomic structure of your bones is as unique to you as your fingerprints. Although skeletons may resemble inert sticks and stones, they are far from it. Your bones are alive, and they respond dynamically to your surroundings and to your lifestyle.

Have you ever fractured a bone? The pain of such a break shows that sensitive nerves are embedded in it.

Did it eventually heal? If your bones were merely lifeless rocks, any breakage would be permanent.

And haven't most of the twists, impacts, and other stresses of daily life caused little or no apparent damage to your skeleton? If your bones consisted only of a simple cement like, say, bone china (a mix of siliceous minerals and powdered bone ash), they would shatter too easily. Clearly there is more to your hidden hard parts than meets the eye.

Approximately two-thirds of a typical bone's dry fat-free mass, representing roughly 3 to 5 percent of an average adult's total weight, is stony apatite which normally lies buried out of sight inside you. We mainly notice our bones when we damage them, or when we have an X-ray taken, and imagining all 206 of them arranged in a complete skeleton is difficult to do without imagining death, too. Unlike your teeth, your bones are not normally seen in the open unless something terrible has happened to you. Nonetheless, imagine zooming in to take a closer look at this remarkable substance upon which you hang your flesh.

The midsection of your index finger is as good a place as any to begin. You can feel the bone in it if you press another finger hard enough against the flesh that covers it. If you were to remove that bone—again, preferably in your imagination—and hold it in the palm of your hand, it would superficially resemble a cream-colored twig rather than a brown one like the Turkana fossil.

In a finger fossil, mineral atoms have worked their way into the bone and altered its color and density. Until recently paleontologists assumed that this transformation happened by strictly chemical means, but new research shows that bacteria may also do much of the work. A study that appeared in *Palaios* in 2010 involved the placement of cow bone samples in river sand with antibacterial agents being added to some of them but not to others. After three months, the bone from sand that contained bacteria was already largely mineralized, unlike the samples in the sterile settings. In combination with careful analyses of the Turkana bones, this new view of fossilization suggests what probably happened more than a million years ago on the shores of Lake Turkana.

Although we can't be sure what killed the boy, his bones and teeth reveal his approximate age as well as his sex and stature. Incompletely developed knobs on the ends of the limb bones and partially unerupted teeth point to an age between seven and fifteen years. The broad architecture of his skull identifies him as a male, and his arm and leg bones put his height close to five feet four inches. Rather tall for a young boy, perhaps, but this was no human. Pelvic features indicate that the maternal birth canal of *Homo erectus* would have been too small to accommodate the larger skull of a modern infant.

The bones were not chewed, so no predator made a meal of the Turkana boy. A crushed scapula suggests trampling by one or more large animals—perhaps a hippo, judging by the presence of fish bones and other evidence of shallow waters. The positions of the jumbled remains imply that the body lay face down in the water and began to decay as fine silt buried it. Bacteria would have consumed the collagen, cells, and blood in the skeleton, leaving behind mineral cement in the spaces that the organic materials once occupied. Iron atoms from hemoglobin combined with sulfur atoms from various proteins, entombing some of the metabolizing bacteria in tiny particles of pyrite. Other iron atoms from the bone capillaries and from wet sediments stained the remains with rust.

Within years to millennia, the formerly porous bones resembled dense, brown ceramic, and if you dared to try it, as I did with a piece of fossilized hippo vertebra from a nearby site, you could actually hear how different the boy's skeleton has become. When I tapped the hippo fossil against a stone, it produced a musical clink. A piece of fresh bone, still porous and full of organic goo, would clunk instead.

The natural tendency of apatite to exchange atoms with its surroundings also allows your body to edit itself. Your living bones are laced with natural additives that weaken the mineral matrix slightly, and acid-soluble carbonate ions comprise about 7 percent of the dry mass of your skeleton. This makes it easier to repair or reshape your bones when necessary. It also turns them into portable quarries from which calcium and phosphorus can be extracted when food is scarce.

The apatite in the hard enamel that coats your teeth forms larger crystals and is more resistant than bone to wear and chemical etching. There are several reasons for this. Enamel contains fewer carbonate molecules than bone, so it is less soluble in mouth acids. Fluorine atoms snuggle into clefts in the crystal structure, making it more rigid and stable. Additional fluoride in drinking water and toothpaste can strengthen it and replace atoms that cavity-causing acids remove. A thin film of protein drawn from your saliva, called pellicle, shields the surface from chemical attack and prevents dietary calcium and phosphorus from binding to your teeth and caking up like stalactites in a cave. Delicate sheaths of protein also encase each apatite crystal and help to prevent fractures.

These are good features to have because you get only one full set of adult teeth during your lifetime. Your permanent dental enamel is one of the oldest parts of your body, having begun to form inside your upper and lower jaws long before you lost your temporary "baby teeth." Some of it was even deposited while you were still inside your mother's womb.

Your skeleton, on the other hand, has other services to perform, and it is much younger than you are because your cells replace an average of one-tenth of any given sector of bone over the course of a year. The relative softness of your skeletal matrix helps your body to rework the atoms within it in response to your lifestyle. In support of this need for constant maintenance, each of your bones contains nearly as much structural complexity as a skyscraper.

The analogy is fitting because your bones often face threats similar to those faced by tall buildings. A high-rise structure built of nothing but concrete would be too brittle to stand for long. The solution to that problem is to pour the cement around a framework of steel rods that reinforce it against high winds or earth tremors. Your bone apatite is similarly laced with "rebar" supports made of collagen, the same strong, flexible protein that puts the spring in your tendons and ligaments. With the combination of strength and resilience that comes with constructing your skeleton from a mixture of apatite and collagen, you can pull at things, pry them apart, or strike them without shattering your bones.

A close-up of the surface of your finger bone would also show it to be honeycombed with microscopic chambers, passageways, and pipes, and there is so much moisture in the seemingly solid mass of compact bone that 20 to 30 percent of the mass of your skeleton can come from water alone. Being an incompressible liquid, this broadly dispersed integral water probably adds to the shock-resistant nature of your bones, too.

When you were born, much of your skeleton consisted of cartilage rather than apatite. That makes sense because cartilage, consisting mostly of proteins and water, is more flexible than mature bone, and being flexible is helpful when you have to squeeze through a birth canal. A person's skeleton doesn't normally finish hardening up until his or her late teens or early twenties.

The conversion of cartilage to bone employs the masonry skills of millions of cells that extract calcium and phosphorus atoms from the diet and pack them into the spaces between protein fibers. A nursing infant supplies those cells with raw materials from milk, which contains tiny clumps of phosphorus-laden casein and calcium phosphate. These clumps are just the right size to disperse blue light through a process known as Tyndall scattering, which gives skim milk a faint blue tint after the fats are removed. The protein fibers in growing bones provide sites for new apatite crystals to grow, and their orientation mirrors the stress forces that their particular section of bone most commonly encounters.

As the matrix solidifies, the hardworking bone cells find themselves trapped. But this is not so much an imprisonment as a shift to an equally important lifestyle, that of caretakers. Each cell continues to live on in a private chamber that offers many of the comforts of home. Plumbing tunnels bring blood vessels close enough to provide food, water, building materials, and waste disposal, and neurons in passageways between chambers help your bone cells to communicate with one another and with the rest of your body.

Should structural troubles arise in your finger, your bone-dwelling cells are ever ready to respond. If a break occurs, they help to close the gap with an easily mobilized version of apatite called "woven bone" that is later replaced by stronger blends of apatite and collagen that resemble plywood. Twisting and bending distorts the embedded cells and tells them to reinforce their surroundings in anticipation of similar stresses to come. And when a nursing mother's body demands more calcium and phosphorus for milk production, her bone cells can mine those elements from her skeleton to supplement what comes in as food.

The living nature of bone makes your skeleton unique to you, and how you live your life is reflected in the architecture of your bones. The legs of runners, for example, contain customized arrangements of apatite and collagen that help to resist the shock of soles against pavement and the tug of muscles against foot-, shin-, and thighbones. The spongy-looking interiors of bones can also be reconfigured in ways that brace against the impacts of running, and thickening of the walls of long bones can produce greater stiffness. Surfers often develop "surfer's knots" on their shins just below the knee from the pressure of kneeling on hard boards. The bones in the dominant arm of a tennis player tend to become stronger than those in the other arm, and a study published in the *American Journal of Sports Medicine* reported that military recruits who frequently played basketball before joining the service were less likely to suffer tibial fractures during basic training. The process works in reverse, too: Lack of gravity-induced stresses weakens the bones of astronauts in space.

How strange it is, then, that the substances that comprise the bulk of your living skeleton are little different from rocks. You certainly wouldn't call a skeleton a person, but neither is it merely a pile of shapely minerals. Rather, it blurs the boundary between the living and nonliving worlds.

When Aldo Leopold was a young forest ranger shortly after the turn of the twentieth century, he experienced an epiphany in the mountains of eastern Arizona. Writing in *A Sand County Almanac*, a book that has inspired generations of conservationists and environmental scientists since it was published posthumously in 1949, Leopold described what happened after he and a companion fired into a pack of wolves that they had found pursuing a deer:

> We reached the old wolf in time to watch a fierce green fire dying in her eyes. I realized then, and have known ever since, that there was something new to me in those eyes—something known only to her and to the mountain. I was young then, and full of trigger-itch; I thought that because fewer wolves meant more deer, that no wolves would mean hunters' paradise. But after seeing the green fire die, I sensed that neither the wolf nor the mountain agreed with such a view.

The story of the green fire is one of the best-known passages in Leopold's book, and the power of its metaphor suffuses the land ethic that *A Sand County Almanac* inspired. It also represents an atomic relationship between wolves and mountains that likewise connects us to the Earth and to all life on it. That connection is the "green fire" of photosynthesis.

Plants have passed their cool green flames of life from generation to generation for hundreds of millions of years. They also spread their energy and atoms through food webs that link soil to plants, deer, and wolves. In an atomic sense the vital fire in the old wolf's eyes arose from elemental connections to the land she lived on and also to the vegetation that sustained her prey. The same is true for you; in strictly atomic terms you exist because plants do. Backtrack your atoms, and you will find most of them rooted in soil.

Every breath you take harvests oxygen molecules that pines, palms, and petunias have made, along with help from plankton in the oceans. Most of the carbon atoms that comprise almost a quarter of your body weight were wrested from the air by maize, wheat, or their relatives. Even if you never take a bite of bread or salad, you still eat former vegetable atoms in the animal products you consume. Much of your body moisture spent time in leaves and stems or formed when your respiring cells made metabolic water with botanical oxygen. And before artificial fertilizers were invented, your ancestors' proteins and genes were built with nitrogen atoms that were mainly fixed by bacteria in the roots of plants.

Which of your atoms are not formerly botanical? Remove your body water and you become a shriveled mummy. Remove your carbons and nitrogens next, and you are reduced to a few handfuls of ash. But even that remnant is tied to the green flames of plants as well. The iron and salts in your blood were pulled from soils by roots, as were the calcium and phosphorus atoms in your bones.

No matter how self-sufficient you may think you are, you are as dependent upon plants as a child is upon its parents. We may often think of roots as little more than cables that hold trees firm against a storm, but like your bones there is more to them than readily meets the eye. The subterranean tangles of a forest can be just as extensive and busy as the canopy, and they represent a bridge between you and the stony bones of the earth.

The next time you have a chance to look carefully at a mountainside, pay close attention to what grows on it. If you were to do this in Arizona where Aldo Leopold met his wolf, trembling aspens and fragrant pines would conceal much of the local geology, and pastel-colored lichens would splatter the bluffs and boulders like paint. This scene might look peaceful from a distance, but if you could adjust the pace of your life to match that of the vegetation before you, you might catch your breath in shock.

The forest is tearing this landscape to bits.

Notice the cracks and seams in the rock surfaces. Trees, shrubs, and lichens are not merely occupying them like flowers in a planter: They are breaking the rocks open, chewing them up, and spitting the pieces out.

This is how stone becomes dirt, the first step in releasing mineral atoms into the realm of life. Wind, rain, rivers, and ice do much of that work, but plants have been sculpting Earth's crust on an even more impressive scale ever since the first appearance of large forests roughly four hundred million years ago.

Similar techniques have long been used in the granite industry thousands of miles to the east of Arizona, on the coast of Maine. The heavy blocks that became parts of the Brooklyn Bridge and the New York Stock Exchange were broken free of their host bedrock through a team effort between hardy New England quarrymen and cold New England winters during the nineteenth and twentieth centuries.

Although the granite of a curbstone or countertop may seem to be impenetrably tough, the multicolored grains that comprise it are not so much fused together as squeezed together, and the contacts between them can be loosened by moisture. Granite is a petrified porridge of glassy quartz grains, sparkling mica flakes, pink or white feldspars, and various dark mineral flecks. Under the right conditions the loose alliance crumbles more easily than you might expect.

In the Maine quarries, workers forced wedges into natural strain lines in the rock, poured water down the cracks, and allowed it to freeze. As the ice expanded, it widened the gaps and allowed the wedges to be driven in further. The cycle was repeated until a great chunk suddenly broke loose. When plants team up with water in similar fashion, a natural version of a Maine quarrying operation can result.

During warm seasons the fine tips of rootlets push into seams and wedge them open. The probing actions can be almost fingerlike, as evaporation during daytime or drought slims rootlets down by as much as a third, allowing them to reach deeper into crevices before swelling again under moister conditions. The thickening of roots with age can also split boulders; winter ice expands and widens fissures; and the outer rinds of boulders and bedrock shrink and swell with temperature, thereby weakening them even further.

The destruction of granite also proceeds on the atomic scale. The mild carbonic acid that normally exists in rain and snow attacks the atomic structures of mineral grains, breaking them down and carrying calcium, phosphorus, and other elements away in solution. Iron-rich grains corrode and stain the rock: Much of the world's oxygen gas is thought to be immobilized in this manner. Sometimes heavily weathered granite can crumble in your hands because the feldspars have disintegrated into powdery clay, leaving resistant quartz grains unsupported in the pulpy matrix like loose teeth.

What does this have to do with your skeleton? Ultimately the destruction of rocks provides you with the calcium and phosphorus in your bones. Most of the mineral atoms in your body were dug from the earth's crust, and chances are good that plants had a hand—or root—in that process.

These atomic connections between stones, plants, and people often operate in ways that can't be observed easily, but that invisibility conceals some of the most remarkable examples of the interconnectedness of life on Earth. The elemental dust of decomposing rock mingles with and often becomes the ash fractions of plants and animals. It is from atomic soil that the green flame in the eyes of Leopold's wolf arose, and it flickers within your own eyes as well.

Heavily weathered granite in Cameroon, West Africa, crumbling into sand, silt, and clay. The iron, calcium, sodium, and potassium atoms that supply the cells of these hands also came from similar rocks. *Photo by Curt Stager*

Imagine, if you will, kneeling among broken sunbeams on the soft, aromatic floor of a pine grove in another range of mountains, this time in New Hampshire. And as you prepare to explore what lies beneath you, remember that unlike plain old sand, soil is alive. Like your bones, it is a dynamic mixture of minerals, organic matter, and living things in which geology coevolves with biology, and the boundaries between self and community begin to fade.

This particular soil is a layered graveyard from which new life emerges. Last year's needles and twigs mingle in a loose brown carpet that becomes more compact as you probe deeper. Bacteria and fungi unravel the carpet molecule by molecule, atom by atom, and you can see its progressive decay as you work your way down into the older strata.

Scrape aside some of the loose detritus and you'll expose a black, tarry layer that represents the organic downwash of all that recycling, a kind of condensed forest soup stock. Carbon compounds and other once-living substances accumulate there in all sorts of microbial waste. Sandwiched between this dark layer and the sandy grit below is a colorfully banded hybrid zone in which organic acids have bleached the dirt to an ashy-looking quartz white and stained the deeper grains with rust.

Like your bones, a forest soil blends the elements of rocks with the stuff of life. A small pinch of it can contain billions of bacteria that cling to shreds of leaf litter, pack into spaces between particles to feed on the oozings from above, and etch their way into mineral grains just as cavity-causing bacteria corrode holes in your teeth. Mucilage and dead cells from root tips add to the feast, and protozoans lurk amid the gritty soil particles to graze on the bacteria. Nematode worms more delicate than silken threads wriggle after the protozoans and are in turn hunted by mites and insects.

In this miniature world an atom from a wildflower or a flake of mica passes among some of the world's smallest organisms like a coin in an underground economy. But this trade network is merely part of a larger one that not only sustains soil communities but also helps to sustain you. The roots that link plants to the ground also link them to diffuse webs of fungal fibers, and the flow of matter through those underground connections is so closely akin to our own modes of trade that ecologists are beginning to refer to it in economic terms. More than three-quarters of all plant species rely on this "plant-fungal marketplace" for survival, forming elaborate alliances that help them to exploit and share resources. These hidden networks of the soil are called mycorrhizae, which translates from Greek to "fungus roots."

In 2002 a research team led by the environmental scientist Joel Blum published a paper in *Nature* that described a surprising discovery at the Hubbard Brook Experimental Forest in New Hampshire. By using isotopic tracers to follow mineral atoms through the soils and food chains of the mountainous watershed, they noticed something strange. Most of the calcium atoms in their soil samples came from feldspar grains, but most of the calcium in the leaves of the forest was extracted from much rarer apatite grains which represented only a tenth of the calcium supply in the soil. And more surprisingly, the trees were not doing those atomic extractions on their own. Most of the mining was being done by fungal allies.

The mushrooms that you may have noticed in a forest or on a lawn may resemble plants, but in some ways they are more like you instead. Rather than trap sunlight and exhale oxygen, they eat organic matter and exhale carbon dioxide. And unlike the light-harvesting canopy of a tree, the cap of a mushroom is merely a temporary spore-producing structure. The main body of the fungus, called a "mycelium," is a living mist of subterranean threads that feed on the soil itself, and it can sprout any number of fleshy parasols when the time comes to scatter spores. The fungal threads are thinner than cotton fibers, but they can also be so extensive that several miles' worth of them can occupy a single thimbleful of soil.

In 2003 the forest pathologist Brennan Ferguson and his colleagues reported finding a record-breaking *Armillaria* fungus in northeastern Oregon whose enormous mycelium spread beneath 3.7 square miles of woods. "The fungus is visible in the clusters of golden-colored mushrooms occasionally seen in the fall on the forest floor," the researcher Catherine Parks explained in a press release, adding that the mushrooms "represent just the tip of the iceberg in regard to its true size and impact upon the forest."

Sometimes the fibers of a mycelium tap into roots in order to parasitize them, but in the case of mycorrhizae they pump nutrition into them instead. Many different fungi can connect to a single root, and any given mycelium can share life-giving molecules with trees and wildflowers many yards away. In some cases the fungi and roots use chemical signals to enlist one another in the construction of networks that resemble the placental ties between a mother and fetus. Fungal alliances such as these can triple a tree's harvest of phosphorus, and some fungi can even kill and digest soil insects and then feed their atoms to neighboring trees through their roots.

At Hubbard Brook fungal miners penetrate deep into the soil in search of apatite grains. When they find them, they quarry calcium and phosphorus from them with digestive chemicals and enzymes. But these rock eaters don't simply hoard their mineral wealth: They use whatever they don't consume to trade with the plants above them.

Analyzing the foliage of the trees, the investigators found that 95 percent of the calcium in the needles of the spruce and firs had been "sold" to them by apatite-etching fungi. Nearly two-thirds of the calcium in the leaves of maples and ferns came from apatite and the rest was pulled from decaying stems and leaves whose atoms, in turn, had been mined and traded in previous growing seasons.

What might plants have to offer in return that fungi desire? Sweets, mostly. Unlike fungi, the leaves of plants spin energy-rich sugars from sunlight, air, and water, and their roots bring water and minerals in while sending sugars back out into the mycorrhizal marketplace.

Such interactions also provide opportunities for cheating, and free-loaders sometimes infiltrate these communities. The tiny seeds of lady's slipper orchids don't carry the usual starchy meal for their embryos as the seeds of sunflowers or oaks do. They survive only through the generosity of fungi until the seedlings grow large enough to fend for themselves. Ivory-colored Indian Pipe flowers don't even bother to make the green chlorophyll that might identify them to you as plants; instead they mooch their meals from fungal donors without offering anything in return.

Why doesn't cheating destroy the system? Apparently some of these atomic traders can identify and punish freeloaders.

A study published in *Science* by the ecologist Toby Kiers and his colleagues demonstrated this in the lab by allowing fungi to grow in separate containers and then hitching them up to clover plants. When different fungi were given different amounts of phosphorus to grow on, the fungi that had less to offer their plant partners received up to ten times less sugar than their more generous neighbors. Somehow, the plants were keeping track and refusing to trade with the cheaters, even when multiple fungi tapped into the same roots. Likewise, plants that offered fewer rewards to the fungi received less phosphorus in return.

These cooperative systems have much in common with the most productive human societies. Enforcement of the rules is mutual and reward based, and both sides are free to interact with multiple partners and to switch partners at will. In closing, the authors concluded that this represents "a clear, non-human example of how cooperation can be stabilized in a form analogous to a market economy."

Similar plant-fungus partnerships often operate in more domestic settings, too. The flowers on your windowsill may share their soil with mycorrhizal fungi, and your garden vegetables may even communicate with one another through a sort of mycological internet. An article in *PLOS ONE* recently reported that tomato plants can monitor the health of their compatriots by eavesdropping on the fungi among them. When tomatoes were infected with a pathogen, they released signal molecules into the mycelial network that triggered chemical responses in other tomato plants nearby, thus giving them time to boost their immune defenses before being infected themselves. And a 2013 article in *Ecology Letters* showed that bean plants can similarly "warn" their neighbors to prepare for an impending aphid attack by sending molecular messages through shared fungal connections.

Without these subterranean systems of mining and commerce, most plants would be unable to survive on the stingy dribble of mineral atoms that slow weathering can provide on its own. And because a large fraction of the world's vegetation eventually becomes food, the trade in calcium and phosphorus ultimately supports you and the rest of humanity.

But just as mycorrhizae represent two-way exchanges, the flow of atoms between soils and life also runs both ways. Not only do minerals produce living tissues: for billions of years, life has produced distinctive minerals, as well.

Robert Hazen is one of the few living scientists to have a mineral named after him. During the last four decades he has written many articles and books on topics ranging from the origins of life to the growth and evolution of crystals, somehow finding time to juggle joint appointments at the Carnegie Institution and George Mason University with a parallel career as a symphonic trumpeter. As Hazen has probed the intersections of biology and geology while also blending science and the arts in his personal life, it seems fitting that he should also be commemorated in a joint production of the living and nonliving realms.

In 2008 the International Mineralogical Association approved hazenite as the name of a microbially produced compound that was recently found in the alkaline waters of Mono Lake, California. It forms when bacteria pull phosphorus, salts, and water from brine and pack them into tiny crystals. The requisite phosphorus atoms wash into Mono Lake in river runoff and are stranded there among the hungry bacteria as the water that bore them evaporates into the dry desert air.

Hazen has championed a new perspective on the linkages between the earth and the life it sustains. As he wrote in an article for *Scientific American*, "If you think of all the nonliving world as a stage on which life plays out its evolutionary drama, think again. The actors renovate their theater along the way." As surely as rocks supply atoms to skeletons and shells, the atoms released by organisms also produce minerals that could not exist on a lifeless planet. "Sturdy minerals," he continued, "rather than fragile organic remains may provide the most robust and lasting signs of biology."

Hazen reckons that the stony innermost planets of our solar system contained roughly 250 kinds of minerals when they condensed from stardust billions of years ago. Many of these still turn up in meteorites, among them olivine and zircon. Weathering and partial melting of the earth's crust later produced new elemental combinations, and the tally of minerals rose further as igneous rocks developed from primordial magma. But more dramatic changes followed the spread of photosynthetic life through the oceans roughly two billion years ago.

In an aquatic version of the metaphorical green fire, cyanobacteria flooded the seas with dissolved oxygen gas, which rusted formerly dark iron compounds in the sediments. As marine oxygen leaked upward into the atmosphere, rocks and sediments on land rusted, too. Hazen and his colleagues estimate that this Great Oxidation Event led to the evolution of more than 2,500 new minerals that would not otherwise exist in quantity. Among these are ocher and iron ore, the gypsum of drywall and plaster, and about two hundred versions of uranium oxide. As tectonic plates collided and sank into the mantle, they carried with them the remains of marine organisms that melted, dissolved, and vaporized under tremendous heat and pressure. Like carbon atoms that enter a tree from the air and reappear in a mycorrhizal mushroom, marine carbon atoms that later burst back up through the crust as diamonds may once have been as alive as the people who now wear them on their fingers.

As aquatic creatures developed protective hard parts, biominerals began to accumulate in the oceans. Pellets of microbial apatite congealed in marine muck, and the evolution of apatite-rich mouthparts and skeletons in beasts of the early seas helps to explain why your own teeth and bones contain so much calcium phosphate today. But perhaps most spectacular was the sheer volume of biogenic deposits that formed after limestone reefs and shells appeared. During the last half billion years, generation after generation of corals, shellfish, and plankton have heaped so much limy calcite and aragonite on ocean floors that you can see much of it from space. The Himalayas, Alps, and Appalachians are full of ancient marine limestone, Britain's white cliffs of Dover are the remains of chalk-shelled organisms, and the Great Barrier Reef shelters more than sixteen hundred miles of Australian coastline.

The biogenic minerals that accumulated in the oceans also dragged heat-trapping carbon dioxide down for burial, cooling climates of the early Paleozoic era by weakening the greenhouse effect. Later in the Paleozoic, forests pulled even more carbon from the air and locked it underground in hundreds of billions of tons of coal, a flammable biogenic rock that we now unravel back into the air.

The green fire of vegetation that springs from the interface between air and stone speeds the erosion of landscapes by tenfold or more. Writing in *Nature* in 2008, the geologist Philip Allen described the astonishing pace at which the continents erode. More than twenty billion metric tons of rock debris wash into the oceans every year, along with a similar

The calcium atoms in this deer skull came to the animal by way of the plants it ate. The plants got them from soils derived from this limy bedrock near Chazy, NY, which was deposited in an ocean reef community 450 million years ago. After long ages underground, the snails whose remains are still visible in the rock now release their shell-calcium into the bones and teeth of deer via local vegetation. *Photo by Kary Johnson*

mass of dissolved minerals. This large-scale sculpting may also feed back into larger movements of the crust itself. In a recent article for *Earth System Dynamics*, German scientists suggested that roots and fungal fibers grind the continents down so much that Earth's molten mantle churns more vigorously in response to the loss of confining weight and insulation. If this hypothesis is correct, then continents may drift, collide, and quake as much as they do because of the trees and mushrooms that ride upon them.

In light of this tendency for life to alter its surroundings, we should not be too surprised to find ourselves doing it, too. Many of us who are concerned about our impacts on the world speak of an ecological balance that we threaten with our disruptive ways. This concern is often justified, but sometimes imbalances can be normal signs of life on Earth, as well.

Consider the massive production of oxygen by forests, the rapid erosion of landscapes by plants and fungi, and the growth of mighty reefs, all of which change the world profoundly. Plants harness the power of the sun, split water, pump fluids and minerals out of the ground, and sustain fabulously prolific communities. By performing similar feats with our minds, machines, and mineral-rich skeletons, we follow a tradition that antedates our species. But this also raises old questions about whether or not we humans are somehow distinct from the rest of nature. Is the commercial excavation of phosphate rock for fertilizer so different from the fungal mining of apatite at Hubbard Brook? If plants can alter the face of the planet, promote trade in natural resources, and change the chemistry of the atmosphere, then why shouldn't we?

From a purely atomic perspective, one might argue that there is little difference. Like all living things we build and maintain our bodies with the atoms of the earth, and like all living things we ultimately return them to the earth in more or less modified combinations and forms. Perhaps what is most different in our case is our capacity to understand what we do. This adds ethical dimensions to our behavior, although many of the behaviors that we consider to be uniquely based in human ethics also make practical sense for non-human life as well.

As socially aware beings, our ancestors learned to survive through cooperation, but plants and fungi also thrive by sharing resources through their subterranean networks even in the absence of consciousness. Sensible codes of conduct help to stabilize our societies, but similar rules of fair play have also sustained plant-fungus marketplaces through the ages. Whatever your opinion may be on the uniqueness of humanity, however, one thing is certain—our atomic nature makes us as much a part of Earth's ecosystems as any other species. And as for the ethical implications of our actions, Aldo Leopold offered some wise advice on the matter. "A thing is right," he said, "when it tends to preserve the integrity, stability, and beauty of the biotic community. It is wrong when it tends otherwise."

As we harvest resources on increasingly massive scales and flood our increasingly crowded world with our wastes, understanding the elemental cycles of life is becoming as important to us as the evolution of upright posture was to our hominid ancestors. An important but also enjoyable mission it is, therefore, to learn as much as possible about our atomic selves, too.

8

Limits to Growth

The agriculturalist holds in his hand the key to the money chest of the rich and the savings-box of the poor; for political events have not the slightest influence on the natural law which forces man to take into his system, daily, a certain number of ounces of carbon and nitrogen.

—Justus von Liebig

We may be able to substitute nuclear power for coal power, and plastics for wood . . . but for phosphorus there is neither substitute nor replacement. *—Isaac Asimov*

Upper Klamath Lake, Oregon, is a pristine ecosystem that produces a miraculous substance that can cure your ills and feed the world—at least according to Cell Tech, one of the first major marketers of a green slime that can be filtered from the lake's emerald-colored waters. They called their wonder product Super Blue Green Algae, or SBGA, and suggested that it could boost your energy, detoxify your body, and improve the nutrition and well-being of your children.

The highly profitable company later closed up shop following a series of legal actions over accusations of deceptive advertising and a wrongful-death suit filed by the family of Melissa Blake, a user of SBGA products who succumbed to organ failure in 2003. Today, if you want to try a sample of Upper Klamath's miracle crop for yourself, you can still order it from several companies that continue the harvest or from thousands of home-based distributors worldwide. You might think twice, though, after consulting scientifically reliable sources, which warn that such plankton gathered from lakes in the wild can contain pathogenic bacteria, nerve and liver toxins, and other contaminants.

Setting aside the most obvious question—Why eat slime?—another question naturally follows: Why is Upper Klamath Lake so green? Promotional literature that various companies now provide online claims that the microbes grow so profusely because the lake is free of harmful human influences and blessed with exceptionally fertile volcanic sediments. But a deeper sense of the atomic nature of life in such a place yields alternative explanations that are less appealing.

The greening of Upper Klamath Lake is worth examining on the atomic scale, not only because the ecological processes and human interest stories behind it are fascinating but also because they are relevant to our global limits to growth. A central character in those stories is phosphorus, whose tendency to glow in the presence of oxygen gave it the early Greek name for the morning star, Phosphorus the Light Bringer. Its complex ecological relationship to us now makes this element a flashpoint for fiery debates over how many people the planet can support, how we raise our food, and even how we wash our clothing.

During the nineteenth century, a German chemist named Justus von Liebig championed a concept that revolutionized agriculture and blurred distinctions between the living and nonliving worlds.

As a youth, he witnessed terrible famines when the 1815 eruption of Tambora in Indonesia clouded the Northern Hemisphere with volcanic ash and triggered a frigid, crop-killing "year without a summer" in Europe. Later, as a professor at the University of Giessen, he studied ways to produce fertilizer without having to depend upon manure production by livestock. But although Liebig also invented a concentrated meat extract that is now sold in bouillon cubes worldwide, he is most often remembered for his Law of the Minimum, which states that plant growth is limited by whichever nutritional element is in shortest supply.

Part of the appeal of the concept, in addition to its accuracy, was the simple analogy that Liebig used to explain it. He likened the productivity of a field to the water-holding capacity of a wooden barrel whose vertical staves varied in length. The water in Liebig's barrel could rise only to the level of the shortest stave before dribbling out through the gap, just as a crop can grow only until it runs out of its most limiting resource. In the aftermath of the Tambora eruption, for example, sunlight became a limiting factor for the farms of Europe. Likewise, industrial nitrogen fixation now removes limits on world agriculture that were formerly set by the seasonal deposition of fertile silt by river floods and the rate at which farm animals excrete their wastes.

Liebig's pioneering research later provided a theoretical basis for Fritz Haber's development of nitrogen fertilizers, and his Law of the Minimum continues to inform farmers today, explaining why a properly balanced mixture of nitrogen and phosphorus is needed to sustain crops. It also helps to explain the greening of Upper Klamath Lake, the fetid blackening of aquatic sediments, and the toxic reddening of tides along coastlines around the world.

Although much of Liebig's work focused on agricultural nitrogen, later investigations have shown that phosphorus often plays a more crucial role in the productivity of aquatic life. In 1973 the ecologist David Schindler and colleagues strung a plastic barrier-curtain through the middle of Lake 226 in the Experimental Lakes Area (ELA), a remote research station in southern Ontario. On one side of the curtain they added fertilizer rich in carbon and nitrogen. On the other side they added phosphorus with the mix as well. Within days the results of their experiment were so obvious that they could be seen from an airplane: one side of the lake was blue and the other was scummy green, a dramatic demonstration of Liebig's law in action.

In this case phosphorus was the limiting nutrient that had held the plankton in check. Phosphorus is normally hard to come by in lakes because most of it is locked away in rocks and soils or buried out of reach on the lake floor, a shortage that is related to its more general scarcity in the universe as a whole. Phosphorus atoms are said to be thirty million times less numerous in our solar system than hydrogen atoms, and hundreds of times less abundant than the other major elements of life. The planktonic algae that support aquatic food chains today draw their phosphorus from the water in which they float, and when the supply of dissolved phosphorus runs low, so does the supply of plankton.

To you and me the clear blue water of a wilderness lake may appear to be normal as well as inviting. But from the point of view of plankton, it represents a frontier waiting to be exploited. Extra nitrogen and carbon were of little use to the algae of Lake 226 without phosphorus in the mix, as a fireplace may be of little use without adequate fuel, and adding that key ingredient was—in a biological sense—like pouring lighter fluid on smoldering coals. With the necessary elements of life suddenly available in abundance, microbial populations exploded just as human populations have done through the modernization of agriculture.

The implications of Schindler's work on phosphorus played a central role in the banning of phosphate detergents, which had previously washed into rivers and lakes from homes and businesses and triggered nuisance algal blooms much like those in Lake 226. Phosphorus is now clearly targeted as a problematic nutrient in runoff from fertilized lawns, farm fields, and sewage sources, which assists efforts to protect water quality.

The research on Lake 226 is relevant to Upper Klamath Lake in another sense. Schindler found that adding phosphorus to a lake not only stimulates plankton but also favors certain kinds of plankton over others. In the ELA, phosphorus gave an edge to species that could fix their own nitrogen, and a particularly nasty group of nitrogen fixers, the cyanobacteria, often dominated ELA waters that received extra phosphorus.

Cyanobacteria resemble algae, but they belong to an entirely different category of life, as distinct from true algae as rats are from roses. Biologists used to call them blue-green algae until closer studies of their cells placed them instead within the bacterial kingdom. The distinction is important, and not only because putting "bacteria" on a health food label makes for a harder sell. Some cyanobacteria produce nerve and liver toxins that make the waters they inhabit unsafe for drinking or recreation. One of the genera of cyanobacteria in Schindler's treated lakes was *Aphanizomenon*, which, as it happens, is also the main component of the plankton harvested from Upper Klamath Lake.

Eutrophication in action. Left: Lake 226 in the Experimental Lakes Area of Ontario, showing the discoloration of the far end of the lake by phosphorus-stimulated plankton. *Photo by David Schindler*
Right: Swirls of cyanobacteria in Upper Klamath Lake as seen from space, August 2005. *Landsat photo, courtesy of NASA*

Far from being pristine, Upper Klamath Lake represents an extreme case of artificial nutrient enrichment. A sediment core that was collected from the lake floor by U.S. Geological Survey scientists tells the backstory of the blue-green bonanza, and their published analyses differ quite a bit from the advertisements that call Upper Klamath Lake "clean" or a "perfect natural ecosystem." Sediments representing centuries of lake history show that the *Aphanizomenon* blooms reflect human activity in the watershed. Cattle ranching, ditching and draining of surrounding marshes, and settlement along the shorelines since the early 1900s have fertilized the lake, thus removing nutritional restraints on cyanobacteria. In a less commercial context, the lake's planktonic eruption could be called a serious pollution problem rather than a wonder crop.

As an *Aphanizomenon* bloom thickens, dying cells rain down from the sunlit surface and microbial decay consumes much of the oxygen in the water. Low oxygen levels, in turn, release iron atoms from their formerly rusty bondage in the bottom sediments. As the confining lid of iron oxide dissolves, mud-bound phosphorus from rotting plankton escapes into the water and stimulates more growth. More troubling still, a notoriously toxic cyanobacterium called *Microcystis* sometimes shares the phosphorus feast in Upper Klamath Lake with *Aphanizomenon*, which can also produce toxins of its own. *Microcystis* contamination of wild-harvested SBGA products was implicated in the death of Melissa Blake, and skeptics speculate that cyanobacterial toxins such as these might explain some of the tingling sensations, intestinal distress, and vomiting that users frequently interpret as energy boosts and cleansing reactions. More recently, a study by the German toxicologist Alexandra Heussner and her colleagues documented microcystin toxin concentrations above recommended exposure levels for children in the Upper Klamath blue-green products they analyzed, warning that consumers of such products risk "serious acute as well as chronic adverse health effects."

Artificial nutrient enrichment has certainly been good for Upper Klamath cyanobacteria but not for other residents of the lake. Two endangered species of fish that evolved in the region now struggle to breathe in the waters that spawned them. The story of this troubled lake also plays out on a global scale as burgeoning human populations quadruple the flow of phosphorus atoms that normally escape from rocks and sediments. Huge "dead zones" are now common in the Gulf of Mexico as the Mississippi River dumps phosphorus from midwestern farms and cities into increasingly productive waters, threatening shrimp and fish that once sustained coastal communities. The Baltic Sea suffers from similar problems, and newspapers in Denmark and Sweden frequently warn bathers to avoid beaches when anoxic waters smother local shellfish and turn coastlines into black, stinking graveyards.

Phosphorus enrichment, of course, can also be seen as a boon to the organisms that the nutrient is feeding, and when it happens in a garden we usually like the results. "Eutrophication," the Greek-derived formal term that describes this process, simply refers to an ecosystem that is well-nourished, and when scientists say that a lake or sea is eutrophic it is technically supposed to be a neutral word. But cultural eutrophication of a water body by human-driven phosphorus enrichment is, from the perspective of most humans, too much of a good thing, rather like eating to the point of sickness.

We are not the first to fertilize the world in this manner, however. The earth itself also dispenses nutritional elements, and one of the more remarkable examples of Liebig's law has dumped millions of tons of otherwise scarce nutrients on distant regions for thousands of years. In this case the life-giving atoms travel not in polluted water but on turbulent rivers of air.

On January 16, 1832, HMS *Beagle* dropped anchor among the Cape Verde Islands off the parched northwestern coast of Africa. On board the British brig was the young Charles Darwin, who was then only a few weeks into the expedition that would later help to inspire his book, *On the Origin of Species*. In his journal Darwin noted that the hot tropical air was clouded with a strange powder. In it were the glassy shells of tiny algae that scientists today know as diatoms or, as naturalists of the time called them, "infusoria":

> Generally the atmosphere is hazy, and this is caused by the falling of impalpably fine dust, which was found to have slightly injured the astronomical instruments. The morning before we anchored at Porto Praya, I collected a little packet of this brown-coloured fine dust, which appeared to have been filtered from the wind by the gauze of the vane at the masthead. Mr. Lyell has also given me four packets of dust which fell on a vessel a few hundred miles northward of these islands. Professor Ehrenberg finds that this dust consists in great part of

infusoria. . . . The dust falls in such quantities as to dirty everything on board, and to hurt people's eyes; vessels even have run on shore owing to the obscurity of the atmosphere. It has often fallen on ships . . . more than a thousand miles from the coast of Africa.

Two decades later Darwin published his testament to the ancestry that connects all earthly life. But it would take another century and a half for the atomic connections represented by the dust at Cape Verde to be revealed.

Unknown to Darwin, a natural wind tunnel was blowing huge plumes of powdered sediment at him from the Sahara desert, including one exceptionally dusty site a thousand miles away to the east, the Bodélé Depression in what is now the African nation of Chad. A fortuitous conjunction of factors was in play there, with seasonal winds, two mountain ranges that focused the force of the winds between them, and the desiccated floor of a freshwater sea that was nearly the size of present-day California seven thousand years ago. What remains of it now, Lake Chad, is a shriveled vestige of its former self, and the exposed bed of what was once the world's largest lake has become the largest single source of airborne dust in the world. When the wind conditions are right, the Bodélé resembles a handful of flour exposed to a powerful fan.

More than two hundred million tons of mineral grains and diatom shells fly out over the Atlantic from the Sahara each year, and geologists estimate that at least ten vertical feet of sediment have been excavated from the Bodélé alone during the last millennium. As some of the dust settles along the way, it coats ships, clogs equipment, and fascinates scientists. But the atoms in those parched clouds are more than mere curiosities. They also affect food chains in Florida and Brazil, and some of them may even enter your own body through meat, seafood, coffee, or chocolate.

On some days the Bodélé winds lift seven hundred thousand tons of dust a mile or more into the air, and satellites show it crossing the Sahara within less than a week. Most of it falls into the sea along with dust from other arid regions of northern Africa, but as much as fifty-five million tons of it flies all the way across the Atlantic Ocean to rain down on the Amazon Basin every year.

Amazon rain forests are lush and productive, but their soils are not necessarily as fertile as you might expect. Intense weathering washes nutritious elements downstream or binds them up in dense hardpans, and rapid uptake by plants removes much of what remains. In order for the trees to get the extra iron atoms they need for their cytochromes and other molecules, they depend on easterly winds to supply them with iron-bearing dust from Africa.

Cattle ranches on former rain-forest territory receive free mineral fertilizer, as well, so the fodder upon which grass-fed cows graze contains atoms that once floated in an ancient tropical lake. The same dust also lands on coffee and cacao plantations that lie beneath the plume and could therefore pass imported African iron along to you in your next cup of mocha-java.

Most of the flow to the Amazon occurs during winter, but in summer the dust drifts farther north, sprinkling the Caribbean and the southeastern United States with atoms from the Sahara. Geologists suspect that the rust-colored soils of many Caribbean islands accumulated over thousands of years from African fallout. It also carries other passengers, as well, including living locusts, pesticides, allergenic fungal spores, and various harmful microbes. Asthma rates in Barbados and Trinidad are among the highest in the world, and some epidemiologists attribute this to foreign pathogens dropped by the wind.

In the open Atlantic the atomic connections between land, air, and sea run deeper. Here, the iron enrichment shows what happens when a formerly limiting nutrient is added to an ecosystem that would normally be starved for it. Shortages of iron block the growth of plankton in large sectors of the oceans where bottom sediments and silty river mouths are out of reach. But not beneath the Sahara dust cloud. According to some estimates, more than seven million tons of Bodélé iron fall into the ocean every year, more than twenty times the amount of iron contributed by all of the world's rivers combined.

Marine algae consume much of that airborne bounty, but another

Dust plume blowing seaward over the Canary Islands from the western Sahara in November 2006. *NASA image courtesy of Jeff Schmaltz, MODIS Rapid Response Team, Goddard Space Flight Center*

kind of microbe depends even more desperately on African iron. *Tricho-desmium*, also known as "sea sawdust," is a marine cyanobacterium whose reddish-brown color is thought by some scholars to have given the Red Sea its name. Like the *Aphanizomenon* of Upper Klamath Lake, it fixes its own nitrogen from atmospheric gas that dissolves into the sea surface, and it is nitrogen fixation that makes *Trichodesmium* especially sensitive to African dust.

Nitrogenase, the chief nitrogen-capturing enzyme, is one of the most iron-rich of all biomolecules. While your blood hemoglobin carries four iron atoms per molecule, a nitrogenase complex carries thirty-four. The rain of rusty African dust therefore turns the tropical Atlantic into a fertile field in which to grow iron-hungry nitrogen fixers, especially *Trichodesmium*, which is believed to dissolve dust particles in order to release their iron atoms. The ammonium waste released by *Trichodesmium*, in turn, becomes food for other plankton in the same way that nitrogen in commercial fertilizer feeds a lawn, supporting food chains that may bring shrimp, fish, and shellfish to your table. Not all beneficiaries of those nitrogen compounds are so beneficial to us, though. Sometimes, particularly in summer, waters along the east coast of the United States turn the color of blood. This phenomenon, called red tide, occurs when a dangerous form of dinoflagellate alga undergoes a population explosion. Within those cells are nerve toxins so powerful that anyone who consumes seafood tainted by them risks paralysis or death. One of them, saxitoxin, disrupts the sodium channels of neurons and is also occasionally produced in *Aphanizomenon* blooms. Fish and manatees are sometimes overcome by red tides as well, and even during mild outbreaks shellfish beds are closed, swimmers avoid the water, and local economies suffer. The critical trigger for many of these blooms is not iron alone but its combination with the nitrogen fixed by dust-fed *Trichodesmium*.

In your case iron is easier to come by. You aren't anchored in place as Amazon trees are, and you don't have to wait for iron to rain down on you as tropical plankton does. People mine it straight out of the ground and ship it to anyone who wants it, and we can also count on crop plants to extract it from soils for us. Some of us certainly are iron limited, as sufferers from anemia can attest. But for the majority of us, iron is low on our list of nutritional worries. Most diets provide more than we need, and it is so efficiently recycled within the body that only a few milligrams are required to replace what is shed or excreted by an average adult over the course of a day.

If any element is going to hold back the tide of human population growth, it must either be rarer or more difficult to harvest than iron. It should be indispensable to us, and it should also be sought by other life forms, which could compete with us in order to get it. It should not be a mobile gas that disperses easily through the atmosphere, and to top off the list, it ought to be easily wasted and lost through carelessness.

Phosphorus fits that description, and forward-looking experts have already begun to warn of impending shortages of it. You carry about a pound of the element inside of you, mostly in your bones but also in many vital components of your cells. Like oil, phosphorus is a finite resource that, sooner or later, may constrain our numbers. And like any element of life, when overused or abused it can also become an agent of death.

To better understand why phosphorus is so important, a closer look at your atomic self is helpful. Starting with a look in a mirror, you can easily come face to face with the phosphorus in your body simply by studying your own reflection.

When you look at your face in a mirror, most of what you see is skin. The fleshy contours are wrapped in thin sheets of cells like the skin of a grape—or perhaps a raisin, depending on the age and condition of your face. An average adult wears eight to ten pounds of skin overall, and most of its weight comes from water and oils. But what you see most obviously in a mirror are the dried remains of epidermal cells in the outermost layers of your skin. If you could peer down through that thin, flaky barrier into the underlying tissues to examine the membranes surrounding the living cells, you would see that much of what you expose most prominently to the world is your phosphorus.

Strictly speaking, it is not only phosphorus atoms but more complex molecules of phosphate that surround your cells. Phosphate resembles a clump of balloons tethered close to a fist, with four oxygen atoms for balloons and a phosphorus atom for the fist. The strings holding the balloons are covalent chemical bonds, and the whole unit can tie itself to other molecules, including oils or additional phosphates. This flexibility in partners is critical to keeping you alive and recognizable. And sometimes, when life and death are less pressing issues, it also helps to keep your shirts clean.

Remember the eutrophication of waterways due to phosphate detergents? One reason why those detergents are algal fertilizers is because they resemble the phosphate-rich molecules in cell membranes. Take your own membranes, for instance. Every cell in your body is shrink-wrapped in a double layer of phosphates with oily substances sandwiched in the middle. This is not a solid barrier but a semiper-meable, flexible one that forms spontaneously when exposed to wa-ter, a self-assembling feature with an odd molecular structure.

A phospholipid membrane molecule is like many of the strange composite creatures of ancient mythology, the mermaids and cen-taurs of old. To build it you snap a phosphate "head" onto one or more fatty acid "tails" of carbon and hydrogen atoms. The slightly charged heads attract water molecules, but the greasy tails prefer to mingle with their own kind. When you mix enough phospholipids into the wet medium that a cell inhabits, they align themselves into orderly ranks with their heads facing the water on both sides of the mem-brane and their tails packed inside like butter between two slices of bread. The epidermal surfaces of your cheeks, then, are crowds of phosphate-headed lipids sitting on their tails.

Similarities between membrane molecules and detergents also help to explain why the latter can pull greasy stains away from your clothing even though oils don't dissolve easily in water. When tossed into a washtub, the negatively charged phosphate heads cling to the positive ends of the water molecules, and this attraction pulls the de-tergent into solution. The tails trail along behind until they meet fel-low lipids in a dirty piece of clothing. When that happens, a tug-of-war ensues until swirling currents tear the detergents away by the roots along with clumps of crud.

Before the phosphate ban, a rinse cycle might flush your soapy wash water directly into a river where some lucky alga could snatch it up for the sake of those valuable phosphate heads. Scale this up large enough, add other phosphate sources such as municipal sewage, fertil-izers, or eroding sediments, and you get the multicolored travesties of the Baltic Sea, the Gulf of Mexico, and Upper Klamath Lake. Phospho-rus also makes the cells in your own body flourish just as it does the plankton in a body of water, and for similar reasons.

What are your ten trillion trillion phosphorus atoms doing inside you right now? Most of them are supporting the calcium phosphate frameworks of your skeleton, and most of the remainder are trembling in your cell membranes from head to toe. They are often removed from the outermost layer of your skin, however, shortly before the dead cells in it flake away and are lost to your surroundings. Those rescued atoms are shunted inward for use in the tissues beneath them, presumably because phosphorus is too precious to waste in the daily shedding of skin flakes.

Phosphorus atoms are also hard at work deeper inside your cells. Each cell contains a nucleus packed with DNA, all of it laced with phosphorus. These threads are usually too slender to see under a normal microscope, but if you were to fuse the DNA from all of your trillions of cells into one long strand it would stretch far beyond the orbit of Pluto. Each strand twists around a matching strand, forming the famous double helix with weak hydrogen bonds bridging the gap between strands. Yet despite the vast amounts of genetic information that it encodes, your DNA consists of only four kinds of phosphorus-bearing building blocks, called nucleotides.

To read your genetic code, enzymes unravel your DNA to expose the nucleotides before reading them off in ticker-tape fashion, and here is another point at which phosphates become especially useful to you. The strong bonds that form between phosphates stabilize the main frameworks of DNA while the weaker bonds between the strands open and close like a zipper. Without those sturdy phosphate backbones, your genes would be too fragile to read.

One of the nucleotides, called adenosine triphosphate, or ATP, also works in isolation as well as in genes. Its main job in that case is to serve as a chemical battery for the cells that, among other things, move your limbs and tell you what the world looks, feels, and tastes like. The magic of an ATP molecule is in the triple string of phosphates that it carries. Pop the outermost phosphate loose, and a tiny burst of energy is released in a manner that allows work to be performed. Pop it back on and you can carry the re-energized bond anywhere in the cell until it is needed to prime a membrane pump, build a hormone molecule, or perform other services that your cells require.

Huge amounts of work go into running your body, and you break so many ATP molecules simply by breathing, thinking, and pumping blood that a pile of their discarded remains from a single day's activities could weigh roughly as much as you do. It's a good thing that you recycle them so quickly that you need only a few ounces of ATP in your body at any given time. Otherwise Liebig's law might have killed you long ago for lack of sufficient phosphorus. Your mitochondria do most of that recycling for you, harnessing energy from your food to bind that crucial third phosphate back into place on its ATP perch. You breathe oxygen primarily in order to power your ATP factories, and the ATP from those factories powers your breathing.

Other, more numerous elements of life are also important to you: Your body, after all, is mainly a watery bag of hydrogen and oxygen. But your phosphorus is especially valuable because it is often more difficult to obtain. Virtually every membrane-bound, energy-producing cell shares a similar need for this critical element, making it a universal bond between you and the rest of life on Earth.

Exactly where do your phosphorus atoms come from, and how long will those supplies last with billions of us sharing the planet? Of all the atoms of life, phosphorus may be the one most likely to limit the number of humans that can exist on Earth at any given time, although experts disagree over what our most limiting resources will be or exactly when and how they might force global population to level off.

We hear much about "peak oil," for example. Coal, oil, and natural gas were built by plants and plankton over millions of years, and when we burn them we scatter their atoms to the winds along with the energy that held them together. We won't ever run out of their component atoms altogether: carbon and hydrogen are simply too numerous. But we could easily run out of cheap fossil hydrocarbon molecules within a century or so because we use them so much faster than they form.

In theory crop plants or algae can spin more hydrocarbons from air and water to replace what we burn, although cost, logistics, and ethics can complicate efforts to grow such biofuels on a scale large enough to satisfy our needs. Ultimately, however, it is the sheer numbers of atoms available for life that fix the boundaries beyond which Earth-bound populations cannot expand, regardless of our technologies and the resilience of our societies.

To put it most simply, we don't make the atoms that sustain our bodies but merely borrow them from our surroundings. What, then, are our purely *atomic* limits to growth? If we can estimate those, then we can at least define the outermost horizons of possibility to better anticipate the future of humankind.

Set aside for now any theoretical plans for colonizing the moon or Mars. This is not about preserving the human race itself by dispersing like spores, but about the billions of people who must continue to live here on Earth because they could never fit into those imaginary starships. The thrill of space exploration and the technology that permits it easily blind us to one sobering fact: This is still by far the most readily habitable homeland within practical reach of us. Even the least attractive towns on this planet are paradises compared to the brutal emptiness of space and the prisonlike conditions under which astronauts and space settlers would have to live, so as we reach for the stars let's remember not to lose touch with the earth.

To begin this thought experiment, a rough tally of your atoms is useful, although exact figures are impossible to calculate. You continually shed atoms and you are always replacing them with breaths and meals, so the total fluctuates from moment to moment as well as person to person. Not surprisingly, different experts come up with different numbers for the average atomic contents of an average human being. But rather than worry too much about being overly precise, it is worth selecting a reasonable set of numbers simply for illustrative purposes.

An article titled "Nanomedicine," published on the Web site of the Foresight Institute by Robert Freitas in 1998, provides such a dataset, which I have used throughout this book. According to Freitas, you embody a sum of atoms that is literally astronomical. If, for example, you weigh 150 pounds, your atoms outnumber the stars in the visible universe. Written in mathematical shorthand as 7×10^{27} atoms, the number can also be written as seven followed by twenty-seven zeroes, or as 7,000 trillion trillion. The stars, as far as we know, are thousands of times less numerous.

Such huge quantities invite more intuitive ways of describing them. If your atoms were the size of coarse sand grains, you could fill roughly two trillion Olympic-size swimming pools with them. And what might a collection of two trillion pools look like? Were you to try to find out, you'd soon run into trouble locating a place to put them all, because the entire planet would be nearly five times too small.

Another way in which to try to imagine your atoms involves money. If each of your atoms were a million-dollar bill, and if you could spend one of them every second, how long would it take to spend them all? Forget it. You could blow 31.5 trillion dollars in a year that way, or pay off the national debt of the United States (as of late 2013, anyway) within the first six months if you so chose. But by the time you finished spending the rest, 222 trillion years from now, all the stars in the universe would have exhausted their hydrogen fuel and been replaced by black holes, dull dwarf stars, and frigid cinders.

More uplifting imagery, however, comes from the poem by Oliver Wendell Holmes that contains the instruction "Build thee more stately mansions, O my soul." Like a mansion, your body consists of relatively simple raw materials, and although small quantities of the rarest heavy elements that came from the largest supernovas—such as zinc or silver—decorate some of your molecules like sparkles in a fancy tile floor, the dominant kinds of atoms in your body can be counted on the fingers of one hand.

In your personal mansion, hydrogen and oxygen are like the bricks and mortar. An adult who weighs 150 pounds normally carries something close to 90 pounds of water, which accounts for most of your hydrogens and oxygens. But these atoms are so abundant on a planet rich in oceans that they aren't worth considering further in our search for atomic limits.

There is more to a mansion than the sheer bulk of it, of course, and other less abundant materials help to make such buildings into unique homes. Chief among those secondary elements is carbon, which you might think of as the lumber in your building. According to Freitas's numbers, your carbon supply represents nearly a quarter of your body weight, or 35 pounds in a 150-pound adult.

How many adult human bodies could Earth's carbon reservoir produce? There is plenty to go around in the atmosphere alone, with about half a trillion tons of carbon riding the winds in the form of carbon dioxide. Assuming that we could convert all of it into human form, it would supply enough carbon atoms to build about thirty trillion people.

This outlandish situation would of course require that enough of the other elements were available, as well, and that you could fit everyone onto dry land after you built them. The latter requirement, however, is dubious. If you could make thirty trillion people stand upright together with each person taking up only one square meter of space, they would cover North America, Central America, and nearly a third of South America. In such a scenario you would run out of basic living space long before you ran short on carbon, and this scenario only takes the atmosphere into consideration: The oceans and sediments hold many times more than that.

With oxygen, hydrogen, and carbon thus accounted for, anything approaching a realistic limit to population growth must come from elements that together comprise only 4 percent of your body. Here the value of mansions as body analogs becomes clearer. You don't have to limit the supply of major building materials in order to disrupt a construction project: A shortage of small but equally necessary components can stop it just as effectively. In this situation phosphorus atoms can be thought of as the screws, nails, and appliance batteries of your body, and many experts are beginning to worry that the supplies of those items may be running dangerously low. A 2010 article in *Foreign Policy* suggested that a looming peak-phosphorus crisis may be the "gravest natural resource shortage you've never heard of," and a recent article in *New Scientist* called it one of the major "environmental hazards with potentially global implications."

So where are those planetary hardware stores that supply phosphorus atoms to your teeth, membranes, and other body parts? Most of them lie beneath your feet. Although dense, hard basalt and other igneous rocks are mined for phosphorus on occasion, their low ore contents (rarely more than 5 percent) and the difficulty of processing them makes more concentrated sources necessary to supply a hungry world. The richest deposits that we exploit today are recycled from eutrophic marine habitats of the distant past.

Long before dinosaurs appeared on Earth, fungi and plants mined phosphorus from early soils. Runoff washed some of those atoms into the oceans where plankton snatched them up, while others fed life on land. Phosphorus atoms that now reside within you formerly belonged to so many and varied creatures over hundreds of millions of years that even a partial list could boggle the mind. Imagine the phosphatic parts of you soaring, swimming, creeping, or undulating in whatever biological form you wish, and there's a good chance that your biofantasy was once a reality.

If you are a typical American, most of the phosphate fertilizers that produce your veggies and feed the livestock that now feeds you come from ancient ocean deposits in North Carolina or Florida, with a smaller portion coming from Morocco. The phosphorus in your seafood, on the other hand, most likely comes directly from marine food chains, and many of the atoms in your bacon and eggs come from pigs and poultry that fattened on fish-meal supplements. But for simplicity's sake, let's take a closer look at the geological sources on land. Their stories will ultimately lead you back to the sea anyway, as well as to the question of our ultimate limits to growth.

Fossil hunters love the open-pit mine on the sandy coastal plain in Aurora, North Carolina. One of the world's largest phosphate producers employs an enormous dragline excavator there, with a long boom and a bucket the size of a one-car garage. With this impressive contraption, employees of the Potash Corporation of Saskatchewan, also known as PotashCorp, remove tons of sandy, shelly, and limy overburden and dump it into piles that draw swarms of collectors. There in the gray, pitted landscape lie millions of pieces of fossil apatite that once helped whales to swim, sharks to bite, and seabirds to fly. Rock hounds and paleontologists have long come here to plunder the trove of marine relics dating back two to five million years and, in some cases, closer to twenty million years. Among the treasures are petrified whale vertebrae, the serrated, six-inch teeth of *Megalodon* sharks, and delicate wing bones from the ancestors of sharp-billed auks.

But more lucrative treasures lie nearly a hundred feet deeper, though they don't look as interesting in a display case. The primary targets of the miners are dark, shiny nodules of apatite ranging from several inches in diameter down to the size of poppy seeds. Potash-Corp unearths tons of them each year, shipping them off to fertilizer-processing facilities and outlets all over the world, which in turn pass phosphorus atoms on to you in your food.

The phosphorus nodules beneath Aurora represent eutrophication on a massive scale. Their origins have much in common with environmental conditions on the human-impacted coastlines of today, where agricultural and municipal phosphates trigger algal blooms and where climate-driven sea-level rise nibbles at the shore. The Aurora deposits also have much in common with Upper Klamath Lake and the discolored waters of the ELA, although they formed long before the dawn of humankind.

Imagine spreading the green waters of Upper Klamath Lake out along most of the Carolina coast in a belt more than fifty miles wide. That's what happened when the Gulf Stream slid closer to shore and collided with submerged bumps and promontories of the continental margin. The resultant turbulence pumped bottom water into vast upwellings, bringing with it tons of dissolved phosphorus and other nutrients. When the heavily fertilized waters reached the sunlit surface, they ignited a marine population boom whose echoes still reverberate among us today.

Plankton clouded the water, feeding great schools of small creatures that fed larger fish, seals, and birds. Whales gathered to feed and perhaps to breed, sometimes leaving their bones on the bottom among the teeth of huge sharks that preyed upon their young and scavenged their carcasses. Meanwhile, blizzards of dead plankton blanketed the seabed. Buried among the fossils in those decaying drifts were phosphorus atoms that may now reside in your own bones and teeth millions of years later.

Bacteria in the oxygen-poor bottom sediments haunted the plankton graveyard, eating the sunken cells, stripping their carbon and nitrogen atoms away, and leaving dense phosphate nodules behind. These deposits would later become fertilizer for America's wheat and cornfields, the phosphorus in your mother's milk, and the phosphoric acid that puts a tangy zing into your favorite soft drink.

Similar shifts in the Gulf Stream happened dozens of times between eighteen and twenty-two million years ago, probably because of periodic rises in sea level. The timing of the pulses, on the order of one hundred thousand years, matches the rhythm of major climatic cycles and therefore suggests that ice ages played a role in the story. Again and again the melting and surging of polar ice hoisted the surfaces of the oceans, allowing the Gulf Stream to blunder onto continental shelves that were previously too shallow for it. Every return visit triggered more upwelling that laid down more phosphorite. The relatively thin beds of ore that lie near the surface in coastal North Carolina today have been mapped beneath younger sediments all the way out to the edge of the continental shelf, reaching thicknesses of a thousand feet or more offshore.

How does this relate to our atomic limits to growth? Experts disagree on exactly how much phosphate is available for human use on Earth, but the U.S. Geological Survey reports that the United States possesses phosphate rock reserves of about 1.4 billion metric tons (tonnes). With an average ore purity of 30 percent or so, that could presumably yield 420 million tonnes of P_2O_5 molecules, of which 185 million tonnes would be phosphorus atoms. The U.S. reserves alone could theoretically supply the phosphorus atoms of about 370 billion people—more than fifty times Earth's current population. Toss in the other major phosphate reserves in places such as Morocco, which is said to possess fifty billion tonnes, and the numbers approach the absurdity of the preceding carbon-based estimates of human limits.

Why, then, do many scientists warn of an impending phosphate shortage? Depending on which sources you believe, current rates of global production and consumption could leave us without enough affordable phosphate within the next century or less. Not everyone accepts that conclusion, though, and skeptics point to previous claims of imminent depletion of resources that were later proved wrong.

Clearly there are plenty of phosphorus atoms here on Earth to build far more people than we would ever want to share the world with, and dwindling supplies could later make today's low-quality deposits valuable enough to harvest. But there is more to the phosphorus cycle than a direct transfer of atoms from rocks to your ribs. Simply counting the number of potential bodies built does not address the additional need to keep them alive and healthy. Fortunately, the largest factor in such considerations, however, may be one that we can readily control—wastefulness.

Most of the phosphorus that we pull from the ground, as much as 80 to 90 percent of it, never reaches our bodies at all. Some of it goes into growing cotton for clothing rather than food, and at least half of the phosphorus that is sprayed on fields around the world is never absorbed by target crops. Rain washes phosphates into groundwater or rivers before roots can catch them, and sometimes more fertilizer is added than the plants actually need. Phosphorus atoms that make it into corn destined for cattle feed often end up in manure lagoons far from the fields that might, in days past, have recycled them. Food spoilage, uneaten meals, and other losses between farm and fork whittle the atomic tally in our bodies down to a small fraction of what comes out of the ground.

Although these atoms do not disappear from the planet when they escape our grasp, many of them end up in places from which we can't realistically retrieve them, most notably the bottoms of lakes and oceans. Deltas at the mouths of major rivers are now reburying tons of wasted phosphorus atoms that were only recently extracted from geological deposits of similar origin at great effort and expense, storing them beyond easy reach. And on their way from land to water, they can also damage ecosystems that support us. Hundreds of phosphorus-induced dead zones have developed worldwide, and the costs of the related fish kills and other water-quality woes are counted in the billions of dollars.

Social and economic factors also cast the shadow of Liebig's ghost over our future. Most of the world's rock phosphate production is controlled by only three nations; the United States, China, and Morocco. The phosphate resources of these countries therefore represent a powerful source of influence over many other nations that depend almost entirely on imported phosphorus for fertilizer.

Take the heavy losses and damage to the environment, the politics and costs of production and distribution, and the limits of existing technology into account, and the concept of peak phosphorus begins to make more sense. The International Fertilizer Development Center estimates that approximately sixty billion tonnes of concentrated phosphate remain within reasonably easy reach worldwide, and various well-informed sources propose that if mining continues at current rates the remaining deposits could last anywhere between fifty and four hundred years. Experts who argue over the likelihood of finding new deposits or greatly improving methods of extraction may shift that end zone farther out or closer in, but all must agree that some sort of practical limit lies closer to us than do any purely atomic limits.

How we obtain and use phosphorus, in combination with how we live our lives, will determine how long and how well this element can supply our needs. We now boost the global loss of phosphorus from soils to seas by a factor of four, and the human component of the phosphorus cycle is a leaky hose that sprays hard-won atoms into waterways, causing unnecessary harm along the way. One practical task for us to undertake with the aid of new atomic perspectives is to somehow close that loop again.

As important to our future as the number of available phosphorus atoms will be our own personalities and lifestyles. In this new epoch of history, human psychology has become inseparable from geology, ecology, and chemistry. If, for example, primarily vegetarian cultures of the world develop an American-style taste for meat, global demand for phosphorus will skyrocket. And even though astronauts routinely recycle the water molecules of their own urine for drinking, the "yuck factor" alone keeps most of us from doing similar things safely with our phosphatic wastes on a larger scale here on the ground.

With careful planning, we may well be able to turn the human phosphorus pump back into more of a cycle again. And as we seek to do so, awareness of our atomic nature could also help to guide us more effectively toward that more sensible and sustainable future.

9

Fleeting Flesh

*It is far better to grasp the universe as it really is than to
persist in delusion, however satisfying and reassuring.*
—Carl Sagan

I would rather be ashes than dust.   —Jack London

The world is coming to an end, and so are you. Just in case you were
wondering.

The atomic nature of things can be uplifting, as when you first realize that you are star stuff and are breathing argon atoms that Leonardo
da Vinci once exhaled. But as with the evolution of young romance
into less-starry-eyed forms of attachment, when atomic reality truly
begins to sink in to your awareness it can bring mixed feelings. With
creation comes destruction; and with life, death. The birth of your
blood iron destroyed the stars that created it. Between their visits to
Leonardo and to you, some of the same nonjudgmental atoms of the
air also toured the lungs of Adolf Hitler. And the atoms that are now
keeping you alive will eventually abandon you—in fact, many of them
are already leaving as you read this.

You can measure the numbers and kinds of elements in various
pieces of you and conclude that you are a blend of watery hydrogen
and oxygen with a few scoops of soot and dust thrown in. Such a sim-
plistic analysis might lead to the conclusion that you are little more
than intelligent mud, and it doesn't reveal the vital difference between
you and the lifeless lump that you will some day become. How can
you be yourself—your wonderful living, breathing self—and also be a
pile of inanimate atoms?

Unfortunately, if you want satisfying answers to such timeless questions you will have to shoulder much of the burden of finding them yourself, because even the most brilliant and well-equipped minds have not yet nailed them down definitively. But some aspects of atomic reality are now well enough understood to serve as trusty compasses for navigating the squishy depths of perception, mythology, and faith. Simply imagining the migrations of atoms to, through, and from your body can be informative, if not downright mind-bending.

If you are like most of us, then you normally think of your body as a stable, well-defined entity. You expect to see your same old self in the mirror every morning, and people hold you accountable for things that you did in the past. But your atomic self will cease to exist moments from now, and earlier deeds that are attributed to you were actually done by temporary collections of particles that are no longer with you.

It is easy to watch much of this flux of matter in action. You eat, drink, excrete, and breathe, and somehow your body persists because you continue to do those things in balanced proportions. It is more difficult, however, to make meaningful connections between what goes in and what goes out, much less what it does inside you, unless you think in terms of atoms and use appropriate metaphors to explain what is going on. If, for example, you can think of yourself as a river as well as a person, then such things begin to make a different kind of sense.

The next time you visit your favorite river, imagine counting the water molecules in it. How futile! The particles are of course too small and numerous to tally, but the larger problem is that they are gone before you can count them. Flow, after all, is what distinguishes a river from a lake, and yet you give your river a name and recognize it when you see it. "There it is, the mighty Mississippi," you might say, as though it were as changeless as pavement. You don't normally say, "In this brief sliver of time, ten gazillion water molecules are temporarily positioned before me, and in this next moment they have all shifted slightly closer to the ocean so that some are now out of sight and half a zillion new molecules have arrived," and so on. Both statements are arguably true, but the former, simpler version is clearly preferable for practical reasons.

Taking the simpler approach to life served our ancestors well, and it has enabled our species to survive for many thousands of years. But if you are well fed and safe enough to be reading a book such as this, then you can afford to dig deeper into what you really are, if only for the sheer pleasure of it.

As atomic motions cause the waxing and waning of rivers, so too do they produce the transient body that you—whatever "you" means—currently inhabit before its substance flows back into the great global sea of atoms. Consider, then, the particles that are departing your body at this very moment. There is no need to wait for death to scatter you to the winds, waters, and soils of the world. It is already happening.

If you start by noticing the bulk flow of matter into and out of your body and then work your way down the size scale, you can follow demonstrable facts to some surprising conclusions. But as you ponder the two-way transformation between lifeless elements and living tissues, beware: Therein lie the outer limits of sanity.

When you take a sip of water it doesn't just slake your thirst. It literally becomes you. The water that runs down your gullet will, within minutes and without processing of any kind, become some of the dominant fluid in your veins and your flesh. Most of your blood is simply tap water with cells, salts, and organic molecules floating in it. Some of the rubbery squishiness of your earlobe poured out of a bottle or a can just a short time ago. And much of the moisture in your eyes only recently fell from rainclouds.

Your mouth is the portal through which water normally enters your body, but you are quite a leaky vessel. A hydrogen isotope study published in the *British Journal of Sports Medicine* reported that the sedentary men under examination consumed and lost about seven pints of body water per day, with four pints leaving through urine and two or three pints through sweat and breath moisture. Vigorous exercise can boost non-urine water losses to one or two pints per hour.

Now let's see what logic can do with those facts. Nearly two-thirds of your weight comes from water, and your body is an eddy in a stream of that common fluid. Surely the liquid that you slurp from a fountain is not alive, and you don't consider it murder to stomp on a puddle of water. Therefore most of you is not alive at all, nor is it even permanent or unique enough to merit a personal name.

Next let's consider your hair. It is a slow-motion shower of lifeless protein that sprays out of your head at roughly half an inch per month or six inches per year. Each filament is a tangle of carbon and oxygen atoms that are outnumbered two to one by hydrogen and sprinkled with nitrogen along with a dash of sulfur. The atoms in the roots are derived from meals that you ate within the last few days, along with some drinks, metabolic water, and your own recycled cells.

Your fingernails are also full of keratin, and they roll out of your fingertips at three to four millimeters per month, on average. Your toenails grow half as fast but they, too, release atoms from their leading edges as you cut or wear them away, and you also shed millions of microscopic keratinous skin flakes every day. If you could fast-forward a video of your hair, nail, and epidermal growth, you would seem to smolder with skin dust while jungles of protein poured from your hairy parts and curly peels of keratin shot from your fingertips and toes. Even at the usual slow pace, that's a lot of atoms, all of which must be replaced if you are to resemble yourself for very long.

One way of summarizing this unusual time-lapse view of yourself is that you are a walking fountain of carbonated water vapor, liquid water, and protein. Much of it trails off behind you in an invisible mist of exhalations and exfoliations, ending up in the dust bunnies beneath your bed or, should you misbehave badly enough, in the nostrils of a bloodhound. If you ever do become a fugitive of that sort, then perhaps you might attempt a plea of atomic innocence should you be brought before a court of law. "It wasn't I, Your Honor," you can truthfully say (and perhaps with excessive grammatical propriety), thanks to the rapid turnover of matter in your body.

There is a lot more to your body now than there was when you could still fit inside your mother's womb, and that fact alone makes it obvious that most of your body is younger than you are. But imagining yourself as a temporary collection of cells can also make the transitory nature of your body more apparent, too.

A study by the Italian researcher Eva Bianconi and her colleagues recently put the average number of cells in an adult human body at thirty-seven trillion. They vary wildly in shape and size, with diameters ranging from eight microns for a red blood cell and about twenty-two microns for a liver cell to roughly one hundred microns for a mature egg cell (five hundred microns would span a grain of salt). Some of them may last for a few days or weeks before being recycled and replaced, while others may last a lifetime. How can you tell which is which?

One way to estimate the turnover rates of human cells is to measure the amount of carbon-14 in them. During the Cold War, atmospheric testing of thermonuclear weapons turned atmospheric nitrogen atoms into radioactive carbon-14 that still contaminates air and oceans today. Moving into plants as carbon dioxide, the unstable atoms of bomb carbon have worked their way through food chains and lodged in the bodies of everyone on Earth, including you. Since 1963 when above-ground testing was banned, radiocarbon concentrations have declined as carbon-rich organic matter has been buried in ocean sediments, and the change is reflected in our bodies. If there is any bright side to thermonuclear pollution, it may be this shifting concentration of bomb carbon that provides a global isotopic tracer for determining the ages of our cells.

Olaf Bergmann, a cell biologist at the Karolinska Institute in Stockholm, recently coauthored a paper in *Science* that used this technique to document the growth of new cells in heart muscle. His approach resolved a long-standing conflict between experts who believed that the heart is renewed as many as four times during a lifetime and those who believed that we die with essentially the same heart we were born with. By measuring the radiocarbon contents of cardiomyocyte cells from which heart muscle is made, Bergmann's team found that the cardiac tissues of relatively young people who spent their entire lives amid the earth's contaminated carbon reservoir were far more radioactive overall than those of older people who were born before the nuclear tests began. From these and similar findings, it appears that you do continue to form some new heart tissue throughout your life, and that you do so at different rates as you age. According to Bergmann's calculations, you replace about 1 percent of your cardiac muscle cells per year at age twenty and half as many at age seventy-five. Nevertheless, you still keep most such cells with you throughout your adult life.

Similar radiocarbon tracer studies suggest that the average replacement rate of most cells in your body is between seven and ten years, but some cells fall well outside that range. Your heart, for example, is full of connective tissue, blood vessels, and other structures that are replaced more often than your cardiomyocytes. A median annual turnover rate of 18 percent for those components suggests that most of your heart is less than five years old.

In a follow-up paper in *Science*, Bergmann and the biologist Jonas Frisén reported that nerve cells within the olfactory bulb and hippocampus of a human brain are continuously regenerated. This means that when a whiff of something sparks a memory, be it a smoky campfire or a familiar perfume, the neurons that originally encoded those sensations may no longer be with you, and the memories may now be preserved by cells that never experienced them. Most of your other brain cells date back to your infancy, but tracer studies now show that some fresh neurons can also appear within your cerebral cortex, perhaps registering new experiences from day to day.

Cells that line your digestive tract are replaced every few days, which is not surprising considering the abuse they take from stomach acids, bile, and erosion by the passage of food and waste. Work by the physiologist Bernd Lindemann posits a lifespan of about ten days for the taste cells in your mouth, and the dermatologist Gerald Weinstein and his colleagues estimate a mean turnover time of thirty-nine days for skin cells, which spend only a couple of weeks in your outermost layers before flaking off by the hundreds of millions. This continuous shedding gives you a new wrapper of skin once or twice a month and a steady supply of house dust to keep up with.

The lives of your red blood cells are rather "nasty brutish and short." After tumbling through hundreds of miles of aortic rapids and hard-to-squeeze-through capillaries, and after repeatedly swelling and shrinking in thousands of transits through the osmotic jungles of your kidneys, most of them wear out within four months or so and must be replaced by progenitor cells in your spleen and bone marrow. And according to the science journalist Nicholas Wade, the replacement times of three hundred to five hundred days that liver cells enjoy can grow you a whole new liver every year or two.

The Swedish biologist Kirsty Spalding and others have found that your fat-storage cells persist for about a decade, which is good news for people who struggle to lose weight. It was long thought that starvation merely deflates fat cells rather than killing them off, leaving them to fill up again like grocery bags when a dieter tires of feeling hungry. But if you can stick to a healthy regimen for long enough, it seems that you can help to stabilize your weight by outliving some of your fat cells.

Your bones and muscles are constantly remodeled. About 3 percent of the dense outermost layers of your skeleton and up to a quarter of the porous bone in the knobby parts of your limb joints are recycled every year, and experts calculate an average life cycle of a decade or so for your skeleton as a whole. The muscle cells between your ribs live for about fifteen years, according to Nicholas Wade, and the collagen cores of your tendons are essentially permanent once they finish developing during your late teens.

Recent isotopic analyses by researchers in Denmark and Sweden show that the oldest easily identifiable structures in your body are the crystalline lens proteins of your eyes and the enamel of your teeth. If you carry healthy ovaries, then you may also carry thousands to millions of microscopic oocytes that formed while you were still in your mother's womb, making the initial cells of your potential future children nearly as old as you are. And as for tattoos, although younger than you they are permanent because the ink is not cellular and therefore not recycled; it is more like the persistent pebbles in a cornfield than the ephemeral crops of skin.

In sum, your tissues are a mishmash of newborn, persistent, and dying cells, most of which are relatively new. Therefore, whatever you have supposedly done to deserve credit or blame, it really wasn't *you* after all, was it? In this mad worldview the most likely culprits would be your eyes, your teeth, some brain matter, and perhaps the seeds of your unborn children.

Just as your body is older than most of your cells, your cells are older than most of their transient molecules. Because of this your apparent age decreases even more on smaller size scales.

In a much-cited report for the Smithsonian Institution that was published in 1954, the physicist Paul Aebersold wrote that nearly all of your atoms are replaced every year. Referring to some of the earliest radioisotopic studies of human physiology, Aebersold declared that "in a week or two half of the sodium atoms that are now in our bodies will be replaced by other sodium atoms. The case is similar for hydrogen and phosphorus. Even half of the carbon atoms will be gone in a month or two." He then added, "In a year approximately 98 percent of the atoms in us now will be replaced by other atoms that we take in in our air, food, and drink." Other frequently repeated estimates appear in Richard Dawkins's book *The God Delusion* and in Bill Bryson's *A Short History of Nearly Everything*, which suggest that not a single one of your atoms or molecules remains after several years. Such figures capture the general idea of human impermanence, but how accurate are they?

The rapid turnover of water alone makes nearly two-thirds of your body come and go within two or three weeks. But the actual disappearing act is even faster. Your water molecules rapidly trade hydrogen atoms with one another, and they also share atoms with larger molecules in your bones and tissues. In a National Public Radio interview summarized by the correspondent David Kestenbaum, the philosopher Daniel Dennett recently described a comparable situation in terms of a joke about the age of an axe. "There it is in a glass case," he explained, "and [someone says] this is Abe Lincoln's axe. So I say, that's really his axe? And he says, oh yes, but…the head has been replaced twice and the handle three times."

Your cells also rip water molecules apart and attach the pieces to fragments of food molecules during digestion. They produce new water when they rearrange those fragments into person molecules later on, and they turn oxygen gas into metabolic moisture. It is therefore safe to say that some of your present atomic self will cease to exist by the time you draw your next breath.

An average 150-pound person carries about twenty-four pounds of protein, not only in muscles and tendons but also in thousands of other forms. The equivalent of eleven to fourteen ounces are degraded and replaced daily, with the lifespans of various kinds of protein ranging from split seconds to years. For example, nearly half of your muscle protein consists of myosin fibers, of which an average of 1 to 2 percent is replaced each day. The same rate applies to the hemoglobin in your blood, and roughly half of the energy-generating cytochromes in your mitochondria are recycled every four to six days. According to a paper by the physiologist Yves Schutz, the labor of this constant grooming and repair of proteins consumes 20 percent of your energy at rest, costing in caloric terms every fifth pretzel you might munch while watching television.

So what about the aforementioned estimates for your elemental turnover? Your teeth and eyes contain atoms that you were born with, and your bones and tendons contain atoms from your childhood, so the complete turnovers proposed by some authors are exaggerated. About a tenth of your skeleton turns over annually, but with the dry fat-free portions of your bones representing up to 7 percent of your total weight, a fair bit of your mineral mass also persists for more than a year. On the other hand, although some of your cells can last a long time their atomic and molecular components do come and go fairly rapidly: The recycling and reshuffling of water and proteins alone replace more than three-quarters of your atoms every few weeks. Overall an atomic turnover rate that takes into account the stable parts of your teeth, eyes, tendons, and bones does indeed seem to be within the ballpark of Aebersold's estimate of 98 percent per year, albeit a few percentage points lower.

You can argue the exact numbers in any of these estimates, but one thing is clear. You are gone before you know it. The dissipation of your atoms that we typically associate with death is a normal and continuous part of life, and you have been melting away into the elemental pools of the earth ever since you were born.

The flip side of this, of course, is that you must also absorb atoms as rapidly as you lose them if you want to survive for very long. How remarkable it is that your body can collect, process, sort, and use the raw materials of life so reliably. Like a river, you are both a consumer and a source of your surroundings, and those atomic connections extend far beyond your body. Your atoms link you not only figuratively but physically to this planet and its inhabitants as well as to the rest of the solar system, and even to distant galaxies far beyond our own.

On October 15, 1991, something that seemingly shouldn't exist crashed into the Fly's Eye cosmic ray detector at the Dugway Proving Grounds southwest of Salt Lake City, Utah. The unexpected speck was smaller than an atom but it struck with so much energy that the physicist Pierre Sokolsky likened the effect to one "that a lead brick has when you drop it on your toe." The flying particle was several times more energetic than another one that struck a detector in New Mexico nearly thirty years earlier with a punch too powerful to have been thrown by any process that was well known at the time. After the record breaking space bullet struck in 1991, stunned scientists called it the "Oh My God Particle."

The Fly's Eye speck still holds the record for sheer impact, but enough similar cosmic pellets have now been detected that astrophysicists can begin to understand what they are and where they come from. Most are particles that travel almost as fast as a beam of light and originate within any number of several thousand galaxies that float closest to the Milky Way. The reason for this distance limit is that even the fastest projectiles lose energy to collisions with the dust and other obstacles of interstellar space, so anything as energetic as the Oh My God Particle must have traveled no more than a hundred million light-years or so. Many possible sources have been suggested, including the plasma of dying stars and jets of radiation from supermassive black holes.

Ultrahigh energy particles such as these represent the upper end of a spectrum of space shrapnel that scientists have been studying for the last century. Early investigators noticed that something in the air tends to disrupt electrical devices on occasion, and they presumed that radioactivity from the earth's minerals was responsible. However, after hauling electroscopes high into the atmosphere with balloons and positioning them on mountainsides at various elevations, physicists found that the impacts tend to become more numerous the farther you move from the center of the earth. The sources lie beyond this planet, not within it.

Cosmic "rays" of this sort are fast-flying bits of matter and energy that zoom in at us from every direction. Most are protons but they can also include electrons, the ionized nuclei of carbon, iron, and other elements, and nearly massless neutrinos that can penetrate the entire planet and then emerge from the top of your head to continue their long journeys through space without leaving a mark on you. By scanning the skies systematically, astrophysicists find that most of the particle paths are bunched within the plane of the Milky Way, which suggests that they come from our own galactic neighborhood. Many of the bits with the highest energy, however, arrive from all directions equally, meaning that at least some of them come from greater depths of space. But all of these particles attest to the ongoing births, lives, and deaths of stars all around us as well as the mortality and mutability of atoms themselves.

The solar wind and Earth's magnetic field divert most of the fast-flying projectiles before they reach the atmosphere, but not all of them. The physicist Nepal Ramesh and his colleagues have estimated that a thousand cosmic collisions pummel every square meter of the upper atmosphere every minute. What happens to such a missile when, by freakishly minuscule chance, it strikes our planet? If it smacks into an air molecule ten miles or so above you, where most of these collisions occur, it may produce an "air shower" of muons and other subatomic shards that a photodetector can record as a pale fluorescent flash against the dark background of a clear, moonless night. The shrapnel could leave a thin cobweb trail of condensed water droplets that contribute to the growth of a cloud, or it might bore through your skin, causing cellular damage, cancer, or genetic mutations, or simply continuing on to plow deep into the ground beneath you.

The higher the elevation of your hometown, the thinner the shield of air and the greater your radiation dosage. An online calculator posted by the U.S. Environmental Protection Agency suggests that a round-trip jet flight between New York and Los Angeles can expose you to about five millirems of space radiation, equivalent to half the dose of a chest X-ray. According to that source, a denizen of mile-high Denver is zapped twice as often (roughly 45–55 millirems per year) as a more low-lying citizen of San Francisco. The Ramesh study, in contrast, put the muon exposure at a mile altitude closer to one hundred times greater than at sea level. They measured about one hundred muon hits per minute per square centimeter at 4,200 feet elevation in California, which could mean more than a hundred muon strikes per second to your upturned face on a hike through the mountains. Whatever numbers you attach to it, this is a risk that we all live with, a natural background of radiation that can cause damage but can also produce new life forms through genetic mutation, a key factor in the origin and evolution of species. Not only do our atoms come from long-dead stars and the Big Bang at the dawn of the universe: Distant supernovas and black holes bring continual elemental change to life on Earth today, as well.

If you follow space science at all, you have probably heard of cosmic rays before. But what happens to them *after* their associated sky showers and cellular havoc are over? Even in the most dramatic demolitions, matter and energy are neither created nor destroyed but merely transformed, and the ceaseless cosmic storm sprays its wreckage against our fast-moving planet like bug splats on a windshield. What happens to all those splats? Do they remain here among us?

Amazingly, some of your carbon atoms are the relatively recent products of just such impacts. When a space proton strikes the upper atmosphere it may trigger a series of billiard-ball collisions that eventually drive a stray neutron into the nucleus of a nitrogen atom. This transforms the nitrogen into radioactive carbon-14, which soon binds to oxygen and mixes downward as carbon dioxide. Thousands of radiocarbon atoms form in this manner every second within every square yard sector of sky, and once they make it to the ground they can be absorbed by vegetation and inserted into food webs, eventually ending up inside you.

A carbon-14 atom's overweight nucleus is unstable, and it eventually spits out a subatomic fragment and reverts to its original nitrogenous self. One never knows when any individual carbon-14 atom will decay like this, but half of any given mass of radiocarbon will do so within an average of 5,730 years. The change can come at any time. If the C-14 atom is built into one of your DNA strands when it decays, its reversion to nitrogen and the recoil of its particle emission can create an error in the genetic code that, if not corrected, can produce a mutation. Isaac Asimov once estimated that roughly three thousand radiocarbon atoms burst like tiny bombs inside a typical adult body every second.

As an heir to the legacy of nuclear arsenals you also contain more radiocarbon than your pre-nuclear forebears did, and similar contamination among the rest of humankind may have caused untold numbers of extra birth defects and tumors during the last half century. As you go about your business, radioactive carbon atoms are exploding all over your body like echoes of the cosmic ray showers and Cold War bombs that produced them, connecting you to the depths of interstellar space as well as to the dawn of the nuclear age.

Some of the water molecules in your body are also cosmic grenades. Certain collisions with nitrogen can produce tritium, a radioactive form of hydrogen that contains two neutrons alongside the usual proton. Tritium-bearing water is not particularly dangerous if you swim in it or if it wafts through your hair on a muggy day, but as with more abundant radiocarbon the relatively weak emissions of tritium explosions can cause more serious damage inside your body. By Asimov's estimate, an average of three tritium atoms decay within you every second. Meanwhile the collision-damaged nitrogen that is left behind in the atmosphere becomes carbon-12, the regular stable form of the element with which you build your body. Because it is identical to normal carbon there is no way of knowing exactly which parts of you are old-timers and which ones are cosmically conceived newborns, but you can safely assume that you carry some of these shape-shifters within you as well.

Space particles smash into the exposed surfaces of rocks, where they can turn mineral atoms into a mildly radioactive form of chlorine that might later dissolve into groundwater and end up in your salad. So continuous is the bombardment that geologists use the isotopic wreckage in their samples to determine how long glacial boulders have been sitting out in the open—which in turn helps to calibrate the history of ice ages. When impacted stones erode and send their atoms off to the oceans, chlorine from those collisions mingles with the earthly elements, so the salt in your shaker and in your bloodstream may contain the remains of space immigrants. You are not only physically continuous with your home planet: Your atomic neighborhood also includes the entire galaxy and even greater depths of the universe.

Combine this zoo of particles with the ionized nuclei of the solar wind, the radioactive decay of potassium-40 and other unstable atoms in rocks and soils, and the flying debris of comets and asteroids, and you get a world that is shockingly different from the one that most of us think we live in. You and your surroundings are silently popping with cosmic impacts, sizzling with radioactive explosions, and glowing with the energy of far-off stars. Meanwhile, the atmosphere is continually eroding as the solar wind sandblasts it into space. It only persists because Earth's magnetic field shields it and because volcanism and biological activity renew it. Your air supply is therefore protected from the worst of the solar storm, but only for the time being. Several billion years from now the dying sun will unload its remaining energy more fiercely, stripping the earth of its atmosphere and then destroying this planet entirely.

With so much change occurring on so many levels, nothing of substance persists forever. And when the flow of matter into your body eventually slows or ceases for long enough, the ephemeral atomic gathering that is you becomes a one-way migration out of town and into the world at large. People, civilizations, species, and even planets and stars must eventually perish, leaving only their atoms to face some ultimate end in yet more profound depths of the future. But in the shorter term between now and eternity what, exactly, will happen to your own atoms when you die?

"Earth to earth," the text of the Anglican burial service reads, and "ashes to ashes, dust to dust." These famous words refer to the biblical passages "Dust thou art, and unto dust thou shalt return" and "I will bring thee to ashes upon the earth in the sight of all them that behold thee." Such sayings are meant to comfort, not to provide lessons in chemistry and biology, and the people who wrote them knew nothing of carbon atoms and covalent bonds, nor did they need to in order to know that we are made of inanimate matter.

You can debate theology and philosophy as much as you like, but there is not and may never be universal agreement on what, if anything, a soul is or what happens to "you" when you die. There are, however, some firm scientific facts about what happens to your body on the atomic level that you can trust enough to work into the foundations of your beliefs and traditions.

In simplest terms most of your body eventually becomes air rather than dust, an invisible gaseous "spirit" of carbon dioxide and water vapor in the sense of the Latin word for breath, *spiritus*. But before we proceed further, it is worth pausing to consider what your atomic nature means for the third to one-half of all Americans who believe in more metaphysical spirits.

There is no credible evidence to support the common idea that a measurable mass of ethereal soul or ghost essence will leave your body at the moment of death. And a 2007 *Softpedia* article by the science editor Lucian Dorneanu explains why it is that, even if you could somehow manage to return as a classic ghost who can pass through walls, you still could not behave as you might expect. You would not really be able to walk, for example, because your immaterial feet would not interact frictionally with the ground, and you probably could not be seen because you would lack the atoms necessary to reflect light. If your ghost were not made of atomic matter then it would presumably consist of energy instead and therefore could not have the density or inertia to manipulate objects or to move air molecules into the coordinated sound waves of speech or a ghostly moan. And as for the clothing that descriptions of supposed ghost sightings typically include, it seems unlikely that the fabrics one wears would become phantoms as well. No, science cannot disprove the existence of ghosts absolutely, but atomic reality does require that they can't look and act as they do in most movies and horror stories.

Moving briefly to other potentially unsettling topics, if you want to know the full gory details of human decomposition, then you can read articles by forensics experts who are associated with places like the University of Tennessee's "Body Farm," near Knoxville, where corpses are left out in the open for scientific study. Suffice it to say here that your atoms will eventually scatter even if you are embalmed and buried in a sturdy casket. But the process happens more quickly and cleanly in other situations, so for the sake of the squeamish let's take an imaginary tour of your theoretical cremation instead.

When you pass from life to death, your cells will run short of oxygen and begin to suffocate. As a result, your mitochondria will stop producing the ATP that unlocks your muscle fibers and primes the ionic pumps of your nervous system. Nonetheless, if viewed under a microscope your recently deceased cells will still appear to be alive because the Brownian motion of particles in their watery interiors will continue to make them seethe and churn. On the atomic scale, movement itself does not necessarily indicate life.

The thermal dance of atoms at 1400 to 1800°F in the furnace of a modern crematorium, however, moves things forward more rapidly. Within minutes the water molecules in your body fluids will be jiggling too vigorously to be restrained by their hydrogen bonds any longer. As they spring apart from one another, they will escape into the atmosphere in vapor form.

As your water departs and your dehydrating remains heat up further, your increasingly restless carbon atoms will also become more willing to leave. But they will need help in order to do so. As your proteins and other organic components disintegrate into soot, the blackened particles will begin to glow with the incandescent energy of breaking bonds and thermal trembling. Airborne "angels" of oxygen will swoop in to gather up your carbon atoms one by one, winging them into the sky as CO_2. The added mass of those oxygen escorts will more than triple the combined weight of your carbon, which for an average adult would yield more than a hundred pounds of thinly dispersed carbonic gas. In similar fashion hydrogen, sulfur, and nitrogen atoms will also escape from your proteins and body fat and fly off in the form of water vapor and gaseous oxides.

Going up the chimney with those gases will be other reshufflings of your late atomic self. Many of the chlorine atoms in the acid of your stomach will once again form corrosive fumes like the ones that carried them out of volcanic vents long ago. If you die with traditional fillings in your teeth, mercury atoms in the amalgam will shake loose and jitter away as well, we hope to be trapped in a filter before escaping to wreak havoc on somebody else's cells. It is indeed possible for the mercury vapor from your fillings to haunt the living in this manner if you pass away in the United States, because no comprehensive federal regulations are currently in place to control such releases. Reliable numbers are hard to come by, but estimates of mercury emissions from American crematoria range from several hundred pounds to several tons per year.

After a couple of hours the furnace will have sorted your former air atoms from the ones that came to you from rocks and soils. All that will be left will be the crumbling mineral matrices of your bones minus their combustible collagen rebar, and a small powdery remnant of salts and rust. Unless, that is, you also carry other items inside you. Despite the best efforts of funerary professionals to discreetly deal with such things, cremations have occasionally been punctuated by the explosion of a pacemaker or silicone implant, and everything from shotgun pellets to metal bone pins and staples may turn up on the furnace floor after the roasting is completed.

When the crematorium delivers your four to seven pounds of flame-resistant leftovers to the people whom you will leave behind, those crumbs won't really be ashes of the sort that accumulate in a wood-stove or campfire pit. Wood ash consists mostly of calcium-potassium carbonates (under relatively low heat) or calcium-magnesium oxides (higher heat). Your own cinders, in contrast, will consist mostly of calcium phosphate from your bones along with a dusting of sodium and potassium salts. The bone rubble in your funerary urn will be fine-grained not because it is ash but because it was crushed to sand size for ease of handling and, presumably, to protect those with delicate sensibilities from the sight of larger chunks.

So much for your body. Most of your atoms will drift off into the atmosphere, and the rest will go wherever the caretakers of your "cremains" send them. End of story? Hardly. When it comes to atoms, your death will not be the end but just another turn of the page in their own epic tales that began billions of years ago and will continue for long ages to come.

After your water and carbon dioxide fumes emerge from the crematorium, they will go wherever the winds take them. If you are burned in the middle latitudes, downwind will most likely be eastward, and if you meet a tropical end it will most likely be westward. This geographic information, along with the work of atmospheric scientists, will allow us to imagine following your atoms as they begin their postmortem journey.

Let's say, for example, that you are cremated in Trenton, New Jersey, as Albert Einstein was in 1955. Like Einstein's vaporized remains, yours will most likely blow out over the North Atlantic, spreading like a plume of smoke as they go. Their exact routes and rates of travel will depend on weather and altitudes, but if you are willing to settle for general estimates you can make some excellent guesses as to what happens next.

Average wind speeds in the lowest ten miles of the atmosphere, where most air molecules gather, typically vary between 5 and 10 mph at ground level and increase markedly with altitude. With 50 mph as a reasonable estimate for a mean speed and 19,000 miles as the circumference of the earth at the latitude of New Jersey, your water and CO_2 will be likely to circle the Northern Hemisphere within sixteen days, or maybe something closer to three weeks in order to account for meandering in the wind belts. During that first flight around the globe most of them will not come into contact with water bodies or plants that might absorb them, so anybody who wants to greet some of your dissipated carbon atoms after your cremation need only to wait a few weeks in order to wave as your rarefied remains pass overhead. However, with typical atmospheric residence times of less than two weeks for water vapor, most of your water molecules may already have condensed into rain, snow, or cloud droplets along the way.

How many of your gasified atoms will still be drifting past an observer after three trips around the world? The surface area of the Northern Hemisphere is about 98 million square miles, so if your carbon atoms mix evenly over it by then, there will be 10 million trillion of them floating over any given square mile at any given moment, or 360 billion of them in a square foot column directly overhead. Therefore, if people were to look straight up at the sky a couple of months after your cremation, they could potentially scan several trillion of your invisible carbon atoms. If they were to wait a few more weeks for bacteria to convert some of your oxidized nitrogen atoms into light-scattering nitrogen gas, they could also rightly imagine you adding a touch of blue to the sky as well.

What will happen to your gasified molecules after that? Their atoms may eventually be woven into wood or released as oxygen gas in some forest, only to revert to metabolic water again in the body of one of your descendants. Some of them will be shattered by cosmic and solar radiation. And some of the highest-flying hydrogens might be pushed into space by the solar wind, where they might ionize, accelerate, and in a sense return fire at the cosmic ray snipers in far-off galaxies.

Within a matter of weeks, most of your remaining water molecules will condense and drop out in precipitation. If they fall into the sea, they will probably trade pieces of themselves with their neighbors and cease to exist in their present forms. However, most of their hydrogen and oxygen atoms will continue to play "musical water molecules" until algae or cyanobacteria split them or they evaporate and fall on land somewhere. Over the next millions of years they will wander the world in countless forms ranging from berry juice to turtle tears.

As for your flying carbon atoms, they will eventually dissolve in raindrops and fall to land and sea, or be snatched from the air by vegetation. Those captured on land will likely be shorn of their oxygen wings and wrapped into the organic molecules of plants, then animals, and possibly some of your descendants. If, as is more likely, they fall at sea, algae will probably grab them. Many of those seafarers will eventually feed fish, whales, and birds, and bacteria will treasure them when they emerge in excreta and skin sheds. In the farther future, some will sink to the seafloor and be buried in mud. Millions of years later, they may ride a subducting slab of ocean crust as it slides down into the mantle where, under the tremendous heat and pressure of those depths, some of them might crystallize into diamonds.

If you believe the advertisements online, some of your carbons might actually become diamonds much sooner than that. If the members of your family so desire, they can hire any number of companies that offer to compress human ash–carbon into an artificial diamond. The instructions are simple: Send the cremains along with a fee, allow several months for the secret process to be completed, and then wait for a lovely memorial gem to arrive in the mail. If your cremation involved relatively low temperatures, there might indeed be enough carbon left in your bones and char to make a diamond. But skeptical bloggers point to the scarcity of carbon in thoroughly scorched cremains and discuss how easy it would be to take your loved ones' money in their most vulnerable hour, wait awhile, and then send a cheap store-bought diamond in return.

If you really want to leave a beautiful carbon memento behind—well, you already have, and it didn't cost you a cent. You have been shedding carbon throughout your life in breaths, skin flakes, and other emissions, and its cumulative mass far outweighs your current body. Your carbon atoms have already spread all over the world, spending time in feathers, fins, flower petals—and yes, let's face it, feces—of every description since your birth. Virtually any place your loved ones go will probably bring them close to some of your former carbons, and some may even reside within their own bodies as well. But if you want to identify more concentrated, specific, and stable deposits to mark your passing, tree trunks are a good choice.

If a tree grows close to a place where you have lived for a long time, then it is likely to have inhaled some of your breath carbons during past growing seasons and woven them into the wood of its trunk. It is even possible to identify which portions of the trunk contain your old exhalations because most trees produce wood in concentric annual rings. The more recently you lived near a tree in your neighborhood or passed it while walking a favorite forest path, the closer to the bark your carbon atoms lie. The preservation of human-associated atoms in tree trunks is not mystical speculation but a fact that is confirmed by isotopic analyses of growth rings in trees that live within and downwind of the CO_2 domes of major cities, by the presence of bomb carbon in tree rings worldwide, and by the analogous incorporation of salmon nitrogen into Pacific Coast vegetation.

The dynamic balance of carbon isotopes in your tissues also shows that you are always dispersing into your surroundings while you live. Because your carbon atoms are continuously replaced through meals that contain less and less carbon-13 and carbon-14 over time, your own isotopic balance has changed during your lifetime as you lost countless former selves. One of the main differences between cremation and your near-total loss of atoms over the course of a year is that cremation disperses you all at once, and perhaps the largest difference between your living body and your mortal remains is not the departure of your atoms but the cessation of new inputs.

Now what about those so-called ashes of yours? It all depends on where they go first. If someone sprinkles them on a field or a garden, your mineral elements will likely soak into the roots of plants. In Einstein's case, however, rumor has it that family and friends scattered his cremains in a river near Trenton, so the Atlantic was his probable destination. Turbulence and mild acidity in the river helped the

Each woody tree ring contains carbon atoms from CO_2 that the foliage inhaled during successive growing seasons. If you had spent a lot of time near this tree, some of your own former carbon atoms would probably be embedded in these rings. *Photo by Curt Stager*

grains to dissolve, and when his atoms dispersed into the sea, algae used his phosphorus and iron for their membranes and cytochromes, and clams and snails cemented his calcium into their shells. Any lingering chips were buried in sediments that may later become phosphorite deposits like those on the Carolina coast, but perhaps the Gulf Stream also swept a few of Einstein's dental phosphorus atoms back toward his birthplace, where they had first emerged from their parent minerals in the farm fields of Germany.

It can be difficult to think in so much detail about death, and it can be even more difficult to accept that the processes described here are not just morbid fantasy but demonstrable reality. But good science deals in facts no matter what we might prefer to believe instead, and if anything can be said to be universally true, you can bet that death is coming for your atoms. So, too, are countless living beings that will populate the future. In light of this, one naturally wonders—how deep might that future be, and does anything last forever?

One rather embarrassing aspect of human psychology is the comfort that we tend to gain from knowing that we are not alone in our mortality. According to the World Health Organization, roughly 58 million people died in 2005, two-thirds of them as a result of old age, and as world population increases, so will annual fatalities. Yes, there is no escaping death, but knowing that it happens to everybody can help to blunt some of its sting.

It also helps to know that not being alive is normal, even for you. Think about it: Where were you when the Declaration of Independence was signed in Philadelphia, or when the Roman Empire fell? You have already been "dead" for all but the most recent slice of the last 13.8 billion years. Statistically speaking, then, it is far stranger to be alive.

Can you escape mortality by producing children to carry your legacy forward? Not for long. On average a daughter or son shares only half of your genes with you, a grandchild carries only a quarter, and so on down the line, and to say that a child "carries your genes" at all is merely a figure of speech. You donate only half of a single set of your chromosomes to a fertilized egg, and the trillions of sets of genes in any one of your offspring are only molecular photocopies of the original set in the zygote that spawned them. And don't forget the cosmic rays, radioactivity, and other mutagens. Genes change over time, which is why you are you and not an amoeba.

Life will go on and on without you, but not forever. Run down the list of things that you think of as permanent, and an atomic view will show that none are truly eternal.

Imagine what will likely happen in the world after you are gone. Hundreds of millions of years of life, death, and change still lie ahead. Every familiar landscape will eventually disappear through uplift, subsidence, and erosion. Continents will collide, oceans will close and reopen, and new mountains will come and go, as will species.

Most of the atoms in the faces, the places, and the creatures that you see around you now will continue to wander among the living and the dead in these future versions of the world. But even this seemingly immortal planet and the atoms that make it up are ultimately doomed in the deeper future.

The sun that now sustains us will eventually swell into a red giant star. As it expands, the thermal dance of atoms on Earth will speed up in response. A billion years from now the sun will be about 10 percent brighter, causing the thermal motions of water molecules to be too vigorous for condensation to occur as readily as it does today. Without sufficient precipitation to sustain them, the oceans will gradually evaporate into the increasingly hot, humid atmosphere. Three billion years or so from now, the sun will be even brighter and water will exist here only in vapor form as it now does on Venus. Plankton will no longer produce oxygen gas, and plant life will perish for lack of rain, from heat, and from the shade of thickening clouds. Animal species will also wink out as the primary suppliers of their atomic nutrients vanish, leaving only subterranean bacteria to await the day when conditions become impossible for any earthly life to exist at all.

When that day comes, a graphical depiction of life's tenure on this planet will resemble a conical lump on the line of history, with a thin leading edge of primitive bacterial diversity beginning about four billion years ago, a peak of total species diversity lasting two billion years or so, and a long wasting away of microbial remnants. Your own life lies about midway through that "lump of life." Long after your brief flicker of existence is over, however, a more decisive end will come five or six billion years from now when the dying sun finally blows your atoms—along with mine and everyone else's—off into space.

But even then, the nuclei of most of your atoms will continue to exist out there long after the dead kernel of the sun cools and the earth is utterly dispersed. They will spread through interstellar space in a nebula much like the one they knew when the sun and planets were born. Many will lose their electrons, but many others will linger in floating particles of dust. For your hydrogens their time on Earth will then have represented about half of their entire existence. Billions of years later some of them might be drawn into stars elsewhere in the galaxy, as might some of your other elements. That would spell the end of them if those stars are large enough to fuse them, although even then most of their subatomic particles would persist in new nuclear combinations.

And this, too, shall pass. Eventually the universe will run too low on hydrogen to launch stellar fusion reactions. The last stars will burn out one by one, and the universe will slowly go dark. Cosmologists place our time on Earth within the Stelliferous Era, a grand fireworks display that began with the Big Bang and that still peppers us with cosmic shrapnel. But even this long era will represent only a small fraction of the total expanse of time. As the universe continues to expand after the last star fuel is consumed, the spreading of the gaps between particles will leave whatever remains of your atoms isolated and adrift in frigid darkness.

If the latest thinking among many cosmologists is correct, then trillions upon trillions of years from now will come a time when all remaining atoms cease their thermal dances and entropy brings an ultimate end, the heat death of the universe. As their energy dissipates, the atoms that now comprise your body—yes, this very finger of yours as well as everything else—will melt into relativistic fogs of subatomic strings and quarks.

So you see, oblivion awaits not only you, but also the atoms from which you are made. And yet this inevitable stripping away of the physical substance of yourself and your world also reveals other truths that, while seemingly contradictory, may offer some comfort as well.

Nothing lasts forever, but you will always exist.

Your body and every other physical object in your world consist of atoms. Such material "things" float in space-time, which Einstein described as a four-dimensional matrix of existence. When matter and energy disperse in the senescence of the universe, no thing will remain and no thing will happen. Only no-thing-ness will last forever, suspended in featureless space-time at the ultimate end of all things.

But if you crave immortality you can perhaps take heart from

Einstein's physics. Positions in time are as real as positions in space, even if you visit them only briefly in your one-way journey through history. The moments of your life remain permanently fixed in space-time, as do the lives and experiences of everyone and everything else for as long as the universe exists.

Although it may sometimes seem that hard-nosed science is incompatible with emotion and meaning on a personal level, nothing could be further from the truth. If your worldview is compatible with physical reality, then even seemingly dire insights can sometimes bring inspiration or a measure of comfort as well. If you doubt my word on this, then perhaps Aaron Freeman might convince you.

Freeman is not an "egghead scientist" but a radio host, comedian, and author, so his human cred is solid. Recently one of his online articles spread widely among scientists and regular folks alike, and I would like to share some of it here with you.

"You want a physicist to speak at your funeral," he wrote:

> You want the physicist to talk to your grieving family about the conservation of energy, so they will understand that your energy has not died. You want the physicist to remind your sobbing mother about the first law of thermodynamics; that no energy gets created in the universe, and none is destroyed. You want your mother to know that all your energy, every vibration, every Btu of heat, every wave of every particle that was her beloved child remains with her in this world.

Freeman continued in this manner, pointing out that "all the photons that ever bounced off your face, all the particles whose paths were interrupted by your smile, by the touch of your hair, hundreds of trillions of particles, have raced off like children, their ways forever changed by you."

He explained that your loved ones need not rely on faith alone in this regard, because the science behind it is accurate, verifiable, and consistent.

And then he concluded with the hope that your family will "be comforted to know your energy's still around. According to the Law of Conservation of Energy, not a bit of you is gone; you're just less orderly."

Such insights may or may not be soothing when the cruel reality of death strikes you or those you love. But at least they can help to make some rational sense of that reality. Death reveals our atomic selves as we truly are, a small but precious part of an ancient, wondrous universe of particles that we so fleetingly appear in the midst of, shape our bodies and experiences from, and—sooner or later—must dissolve into.

Epilogue: Einstein's Adirondacks

*When the breath freezes into ice dust and falls almost
silently to the ground, Siberians call it the whisper
of stars.*   —David Shipler

*The most beautiful experience we can have is the
mysterious—the fundamental emotion which stands at the
cradle of true art and true science.*   —Albert Einstein

Albert Einstein died one year before I was born. Refusing surgery to
address a ruptured aortic aneurysm in April 1955, he said, "It is taste-
less to prolong life artificially. I have done my share; it is time to go."

Although I never met him, I feel a connection to him as many of
us do. He changed the way we think about ourselves and the universe, and although many of his predecessors and contemporaries
also contributed to our knowledge, I like to think of him as the
person who most decisively ushered humankind into the realm of
atoms.

In addition to the intellectual legacy that we all share, however, I
have something else in common with Albert Einstein that I was un-
aware of until I began the background research for this book. We
have both known and loved the Adirondack Mountains of upstate
New York.

Many images of Einstein show him at a desk or posing beside a
chalkboard, but he spent as much time as possible playing classical
violin, sailing, and savoring the outdoors. "Look deep into nature,"
he once said, "and then you will understand everything better." He
also enjoyed crafting vivid thought experiments and garnished his

explanations of relativity theory with tales of speeding trains and slowing clocks. A similar but more experiential investigation into the realm of atoms could instead lead you to Lower Saranac Lake in the northern Adirondacks, where Einstein spent several summers near the end of his life. After all, if atoms are indeed as real as he demonstrated, then the implications of their existence should apply in a lovely wilderness setting as well as in a physicist's lab or an astronomer's distant supernova.

This journey to Einstein's refuge may be purely theoretical for you. But for me the lake and its atomic surroundings will be quite real, because I live just a few miles down the road from them.

Imagine that we leave the four-way intersection beside the town hall in downtown Saranac Lake, cross a short bridge, and turn left to follow the shore of Lake Flower. The historic wood-frame homes are attractive and conveniently close to water, and people still sail the narrows of Lake Flower as Einstein did three-quarters of a century ago.

The place we are looking for lies farther up the road on a wooded hillside, with heavy timbers set in the kind of white plaster that might adorn an old-world chalet. As we approach the front door, a friendly dog sidles up and wants to be petted. The door opens, and the aroma of fresh-baked cookies wafts over us. The woman who introduces herself as Paddy is a skilled baker, and she knows all about Einstein's summer cottage.

"Yes, this is the place. Come on in and look around."

Paddy and her husband, Mike, show us a cigarette burn on the wooden mantel in the dining room. "We're pretty sure it's his," Paddy says. Even if it isn't, atoms from his breath and skin probably mingle with those of the current residents in the pores of the wood.

The living room is long and cozy, with a rustic fireplace and big windows that look out over a descending hillside cloaked in hardwoods. "Back in 1936, Einstein and his wife could see the lake from here," Mike explains. "Now the woods have grown up and blocked the view." Much of this Adirondack landscape lost its trees to axes and ac-

The house in Saranac Lake where Albert and Elsa Einstein summered in 1936.
Photo by Curt Stager

cidental fires during the late 1800s and early 1900s. The forested hills around Saranac Lake are wilder looking now than they were when Einstein knew them, and only the inner trunks of the oldest trees closest to the house are likely to contain much of his former carbon.

Paddy leads us upstairs. "The bedrooms are original, but the bathroom is not." When the bathroom was remodeled several years ago, some of the local boys carried "Einstein's toilet" off and buried it in the woods. "It was their special prize," she says, "and to this day they've kept the burial spot secret."

Paddy believes that the smallest bedroom, situated in a corner facing the road, was Albert's. "He and Elsa kept separate bedrooms. This larger one back here with the big closet was probably hers."

I remember reading about Elsa, and about how she and the man whom she called "Albertle" (the German for "little Albert") married despite their close family ties (they were cousins), and the wild-haired

Einstein at Lake Clear, near Saranac Lake, in the summer of 1936 with Elsa (seated, perhaps reflecting her ill health) and his step-daughter (behind her). *Courtesy of the Adirondack Collection, Saranac Lake Free Library*

celebrity's apparent disregard for marriage vows. Perhaps they both preferred privacy at night. Accounts of Einstein's troubled family relationships suggest as much, although they probably do not reflect everything that went on in the personal lives of this famous couple. Elsa suffered from a long, painful illness, and hopes for rest and recuperation had brought them to Saranac Lake in the summer of 1936 before the malady finally killed her several months later. Standing in their former residence, it is easy to suspect that Elsa's agony may also have made it difficult to share a room at times.

Our hosts assure us that we can stay as long as we like, but it is time for us to move on.

One evening during that first summer, Einstein was invited to dinner at Knollwood, a secluded forest retreat on the shore of Lower Saranac Lake just west of town, where he would spend several more summers after Elsa's passing. It was owned by a club of influential families including that of Robert Marshall, a noted conservationist who helped to found the Wilderness Society. This was the perfect place for a getaway: beautiful views of the mountains, a comfortable cottage among several others nestled close to the main lodge, and a lovely island-studded lake to sail on. Marshall later described his guest's reaction to the setting. "I have seldom seen a person more delighted with the natural scenery than was Professor Einstein," he wrote. "Repeatedly he exclaimed about how different it was in America, where you could still see places which did not indicate the evidences of man."

When we launch our canoe shortly after dawn from Duso's Marina, where Einstein regularly moored his sailboat, and begin to glide across the glassy calm surface of the lake, we are supported by molecules whose behavior Einstein described in his writings on Brownian motion. Hydrogen bonds link these water molecules so tenaciously despite their trembling that the gravitational pull of the planet flattens the whole seething mass of them into a glistening liquid plain. Nonetheless some of them are rising invisibly from the surface, only to chill and reappear in wisps of morning mist, and the fragrant white pines along the far shore are feeding more such vapors to the clouds. As we press our paddles against the thick liquid, the give in the moist flesh of our hands reveals the presence of similar molecules that are also dancing and streaming out of us with every breath.

Reflected in the broad mirror of the lake, the molecular nitrogen above us blazes with the blue tones of the solar wind. It is difficult to imagine that Einstein did not think of his contributions to our understanding of that phenomenon when he, too, floated here between lake and sky. As we paddle closer to Knollwood, airborne nitrogen jostles its atomic relatives in the protein filaments of our hair as if in greeting. When those proteins fall away from us later on, their unraveled nitrogen may mingle with atoms from Einstein's iconic white locks in this same sector of the atmosphere. Later still, those same atoms may be harvested by the process Einstein's friend Fritz Haber perfected and then be passed along to future boaters who will enjoy the wind in their hair as we do now under the nitrogen-blue sky.

Einstein and sailboat on Lake Flower with New York senator John Dunnigan, July 3, 1936. *Courtesy of the Adirondack Collection, Saranac Lake Free Library*

Bull's-eye rings appear on the lake surface around us as a passing cloud trails a brief drizzle of meteoric water. The hissing of these impacts hints at the unseen showers of cosmic particles that also strike us and the lake, as well as other atomic connections that the raindrops represent. Nitric and sulfuric oxides emitted by traffic and coal-fired power plants in the midwestern states melt into the lake, lowering its pH. The acidic rain also feeds nitrogen into the food webs of the surrounding woods as well as the lakes and rivers, turning them into "pollution forest" analogs of the nitrogen-enriched salmon forests of the Pacific Northwest. As a consumer of local berries and trout, I must also carry some of that exhaust nitrogen in the proteins of my hair. However, I prefer to think of it less in terms of pollution in the thinning thatch of my late middle age and more in terms of a grizzled old bear who carries exotic nitrogen in his fur.

A loon surfaces ahead of us with a small perch in its bill, gulps it, and dives again. Mercury atoms from the power plants far upwind are now dissolving out of fish and into bird, perhaps making this loon more sluggish and less likely to tend its nest properly. Oxidized atoms from the fossilized ancestors of this forest tumble between us and the trees, accompanied by CO_2 that their companion fumes displaced from distant oceans. Together they supply the carbon budgets of one in four of the needles on every pine. When Einstein first came here, they supplied roughly one in ten.

When news of the destruction of Hiroshima and Nagasaki reached the Adirondacks in early August 1945, Einstein was vacationing at Knollwood, spending much of his time sailboating with his sister, Maja. In a brutal demonstration of what $E = mc^2$ can mean, the transformation of less than an ounce of uranium and plutonium atoms into pure energy had incinerated tens of thousands of human beings, and the reporter Richard Lewis from the *Albany Times Union* drove to Knollwood on August 11 to ask Einstein for his reactions.

With the help of the club superintendent, he met Einstein that evening at Cottage 6 and recorded an exclusive impromptu interview with the man who had already turned down many other such requests from around the world. After expressing his dismay over the bombing, Einstein told him, "In developing atomic or nuclear energy, science did not draw upon supernatural strength, but merely imitated the reaction of the sun's rays. Atomic power is no more unnatural than when I sail my boat on Saranac Lake."

Einstein did not contribute directly to the development of nuclear weapons. He was a pacifist, and he had also apparently doubted that the splitting of atoms for energy production was attainable. His contemporary Ernest Rutherford had likewise dismissed speculations about nuclear power, saying, "The energy produced by the breaking down of the atom is a very poor kind of thing. Anyone who expects a source of power from transformation of these atoms is talking moonshine." And even many scientists who believed that fission might be

Boathouse and dock at Knollwood on Lower Saranac Lake. Einstein's Cottage 6 residence is hidden among the trees behind them. *Photo by Kary Johnson*

possible worried that it was too dangerous to attempt. In his Nobel Lecture in 1922, the chemist Francis Aston said, "The remote possibility must always be considered that the energy once liberated will be completely uncontrollable and by its intense violence detonate all neighbouring substances. In this event, the whole of the hydrogen on earth might be transformed at once and the success of the experiment published at large to the universe as a new star."

But in 1939 when a friend told Einstein that Germany was developing an atomic super-bomb, he reluctantly signed a letter to President Roosevelt urging him also to do so. He later regretted that decision and strongly advocated against the further development of nuclear weapons. If he had lived longer, he might also have protested the contamination of his body and those of everyone else on Earth with cesium-137, carbon-14, and other radioactive isotopes by the explosion of such devices in the shared planetary pool of air. The muddy floors of Adirondack lakes still contain layers rich in bomb cesium that date to the mid-twentieth century, and in my own research I have also used the same global cesium spike to date sediment core records from lakes as far away from here as Africa.

As we approach the boat dock at Knollwood, the sun burns through the blue depths of the sky. Einstein must surely have marveled at how elegantly his famously simple equation of energy and matter explains that blinding fire. Here in the bright glory of an Adirondack summer day the wave-particle duality and the atomic origins of starlight are on full display. The lake, the islands, and the forests are visible to us only because the visible energy of hydrogen fusion splashes away from them in all directions before striking our eyes. The heat that accompanies the light also orchestrates the thermal dance of atoms here so precisely that water can simultaneously produce vapor, ice, and liquid in this landscape whenever the North Pole tilts slightly away from the sun during winter.

When not considering Einstein in terms of mushroom clouds and scrawled equations, popular culture often associates him with the night sky, presumably because his best-known work dealt with space-time and the light that allows us to look back into history when we view the cosmos. Such talk of space and stars evokes images of standing out under the Milky Way and pondering the black void, but his ideas concerning matter and energy have profound implications even in broad daylight in the Adirondack wilderness.

Simply replacing the word "sun" with "star" can change your sense of what this sylvan scene actually is. Lie flat on your back on the warm wood of a dock, and it may further dispel the normal illusion that the great fireball is "up there in the sky" instead of "right over there beside us in space." Something about being horizontal and seeing the sun-star before you rather than above your head makes it easier to sense the absence of supporting pedestals or cables and therefore to realize that the brilliant, life-sustaining heart of our solar system floats in emptiness as it directs the trembling of your atoms from millions of miles away. So, too, does the rocky globe that now more obviously presses your body against itself, as it also does with the molecules of the lake.

From this position it becomes easier to sense the gently curved surface of an enormous sphere of terrestrial atoms that rolls you and Lower Saranac Lake eastward at nearly the speed of sound. It also becomes clearer that "day" is more of a place rather than a time, a zone in which the solar spotlight draws a blue haze of glowing nitrogen across your view of more distant stars. That veil hides all but the sun-star, the reflective faces of our moon and neighboring planets, or, on rare occasions, a guest supernova such as SN1054. If you wait here long enough, though, you and your patch of Earth will slide out from under the blue mist into a clearer view of the dark wells of space that always surround you in every direction.

If you can grasp this even for a moment, and thereby make your utter dependence on this finite, floating planet feel more tangibly real, then perhaps you will also find it easier to appreciate the atomic reservoirs that sustain you and to understand the effects that you can have on them. It is only here on Earth that life as we know it currently exists. How amazing to exist at all and how important it is, as our numbers and know-how increase, that we and our descendants develop such awareness as best we can.

One can only wonder how Einstein might have wrestled with the still-open question of how inanimate atoms produce life. He freely acknowledged the limitations of human understanding, including his own, and in July 1945, he wrote in a letter from Cottage 6, "We have to admire in humility the beautiful harmony of the structure of this world—as far as we can grasp it. And that is all." The pressures of natural selection have made our brains clever enough to keep us breathing and breeding, but they have not made us omniscient, and there are many things that we simply can't wrap our heads around. Science alone can take us only so far in our efforts to grasp the world, but sometimes teaming it with the arts can carry us the rest of the way forward on that journey. As a musician, Einstein understood this, and perhaps his love of music offered him insights into how life arises from atoms in ways that are now described in terms of "emergence."

An emergent phenomenon arises from relatively simple components that somehow become more than the sum of their parts, as random scratches become letters if they are shaped in certain ways. Letters can be grouped into words with meanings that depend upon their sequences. The letters *e, l, f,* and *i*, for example, can become "file" or "life." Emerging from the same kind of mysterious zone wherein the arrangements of words produce literature, teeming atoms and molecules somehow become our living cells. In similar fashion a thousand minnows produce an undulating shoal of silver, a million citizens make a city with a distinctive identity, billions of coral polyps produce a complex and colorful reef, and trillions of mindless cells create a colony that walks, talks, and thinks of itself as a person.

Music, in this context, is an emergent phenomenon that arises from sound waves in air, and even if it can't completely explain the origins of life it can help to describe life while also making it more enjoyable. Einstein was an excellent violinist who particularly loved Mozart's music, and as his fame spread he was often invited to perform with some of the world's most accomplished musicians. Pianist Artur Balsam, when asked about the musical abilities of the revered author of relativity theory, replied, "He is *relatively* good."

But Einstein's relationship to music was more personal than professional, and although he could have owned the best of instruments he preferred to lug an inexpensive fiddle in a battered case wherever he went. This included Knollwood, where he often played alone on the veranda at Cottage 6 and also enjoyed playing duets with the concert violinist Frances Magnes, another frequent summer visitor to Lower Saranac Lake. "If I were not a physicist," Einstein once said, "I would probably be a musician. . . . I see my life in terms of music."

The science historian Arthur Miller described Einstein's admiration for Mozart as a natural blending of scientific and artistic sensibilities. "The laws of nature," he wrote, "were waiting to be plucked out of the cosmos by someone with a sympathetic ear." In physics as in music, the world contains "pre-established harmony exhibiting stunning symmetries" that Einstein recognized and treasured. To him Mozart's music "seemed to have been ever-present in the universe, waiting to be discovered by the master," which also brings to mind what Michelangelo might have said about his sculptures—that rather than create his figures, he set free the ones that were already present in the stone.

What can music reveal about the atomic nature of life? Physicists sometimes compare the oscillation patterns of orbiting electrons to standing waves in the resonant strings of musical instruments, and the vibration patterns of subatomic superstrings have been said to resemble harmonic sequences that can be played on a violin. But atoms are more difficult to define when considered in terms of quantum mechanics, and both music and life resist precise definitions as well. Biologists still argue over whether some animals' vocalizations constitute song or mere noise, and even scientists who study the origins of life on Earth have no firm definition of what life itself is. Try it yourself sometime, as I do with students in my Introductory Biology class.

After the students list a dozen or so features including eating, respiring, responding to stimuli, and reproducing, I unveil a chain saw that lay hidden behind the lecture podium. As jaws drop and laughter erupts, I pull the cord and the machine roars to "life." Nearly every feature on the list is displayed in the consumption of fuel, the exhalation of waste gases, and the raucous responses to my trigger finger. When I "kill" the engine, someone always asks, "Wait, what about reproduction? It can't be alive if it can't reproduce." And as you might guess, a fairly crude reply soon follows, along the lines of "What about a nun, then? Isn't a nun alive?" or "What about a mule? A mule couldn't reproduce even if it tried."

If defining life is this difficult, no wonder we struggle so much to understand how it arises from atoms. But even if we can't fully explain what life is, the emergence of music from vibrating molecules can help to describe what life is *like*. Consider what might happen if you were to borrow Einstein's violin, which is still played in concert by his great-grandson Paul, and use it to perform one of his favorite melodies, Mozart's Sonata in E Minor, here on the dock at Knollwood.

Most of the atoms of this particular instrument were also here during the 1940s because atoms tend to persist in objects such as violins longer than they do in more transient entities such as lakes and musicians. But what exactly is the music that emerges from your fingering and bowing of the strings?

The sounds themselves are short-lived waves of air molecules striking your eardrums, and your perceptions of pitch and tone emerge from waves of neuronal ions that trigger emergent sensory and emotional responses in your brain. The melody itself, however, is a metaphysical pattern that emerges from the process of playing and, ultimately, from a lyrical thought in Mozart's mind in 1778. The emergent phenomenon of the Sonata in E Minor outlasts any single performance or player, and it exists with or without the instruments that embody it in sound or the scribblings that transcribe it to paper.

Perhaps that is what you are most like, then, not the physical instrument of your atoms but the unique pattern that emerges like music from their interactions, an abstraction that is nonetheless real. Perhaps you are a living melody that successive orchestras of atoms perform in the theater of your body until, sooner or later, the concert series ends. Walt Whitman suggested as much when he wrote:

> I celebrate myself, and sing myself,
> And what I assume you shall assume,
> For every atom belonging to me as good belongs to you.

Like the sound of a sonata, like Mozart, Einstein, and Whitman, you too will be gone someday. But like the abstract structure of a musical composition, the emergent patterns of your life are immortal, and your atoms will continue to exist in many and varied forms for trillions of years until even they must melt into the silence of a dying universe. As Whitman concluded,

> I depart as air, I shake my white locks at the runaway sun,
> I effuse my flesh in eddies, and drift it in lacy jags.
> I bequeath myself to the dirt to grow from the grass I love,
> If you want me again look for me under your boot-soles.

In the meantime, welcome to your atomic self. Hydrogen has become you after billions of years of stellar fusion and countless dances of atoms in air, water, earth, and fire, and may you bequeath them with grace to the many lives yet to come. As you finish the rest of the story of your life, may you share your matter and energy ever more wisely and well with the universe.

Now take another breath, if you please, not only because you must but, wonder of wonders, because you can.

Notes

PROLOGUE

Dimensions of an oxygen atom: The atomic radius of oxygen is 60 pico-meters or 60×10^{-12} m (Slater, 1964), and a reasonable estimate for the radius of an oxygen nucleus is 3 femtometers or 3×10^{-15} m, with the radius of a proton being about 0.84 femtometers (Pohl et al., 2010). Inflating the nucleus to the size of a raspberry with a radius of 1 centimeter would be to increase its size by roughly 3.3×10^{12} times, so the radius of the whole atom would then be $(3.3 \times 10^{12}) \times (60 \times 10^{-12}$ m) or about 200 m. The volume of the spherical atom would then be about 33.5 million m^3 or 43.6 million cubic yards. The volume of MetLife Stadium is said to be 1.8 million m^3, roughly 1/18 of the volume of the inflated atom.

Mass of fingertip made of pure nuclei: Most sources put the density of an atomic nucleus on the order of 10^{17} kg/m^3 (for example, the HyperPhysics Web site at http://hyperphysics.phy-astr.gsu.edu/hbase/nuclear /nucuni.html). The volume of a fingertip is roughly one cubic centimeter, or a millionth of a cubic meter. The mass of a fingertip's worth of nuclei would therefore weigh roughly $10^{17} \times 10^{-6} = 10^{11}$ kg. At 2.2 pounds per kilogram, that converts to 220 billion pounds or 110 million U.S. tons.

CHAPTER 1: FIRES OF LIFE

Calculations for the journey of an oxygen atom down your arm: The atomic radius of oxygen is 60 picometers (60×10^{-12} m; Slater, 1964), so an oxygen atom is 120 picometers or 120×10^{-12} m wide. If the atom's diameter were to inflate 10^{10} times, its diameter would be 1.2 m. For ease of calculations, let that be the height of a (short) person so the scaling factor to use on you will be 10^{10} times. If we also take the approximate length of a typical adult arm to be 0.6 m, it would convert to an equivalent distance of 0.6×10^{10} m, or 60 million km. At 0.62 miles per km, that would be 37 million miles, and to travel that distance in one second would require you to move much faster than the speed of light, which is "only" 186,282 miles per second. Einstein's work on relativity showed this to be impossible.

Dimensions of components of a cell that has been magnified 10 million times to the size of a 300 foot hill: The diameter of a human cell is on the order of 10 microns or 10^{-5} m (Daniels et al., 1979), so magnifying it 10 million times (10^7 times) would inflate it to about 100 m in height (roughly 300 feet). The diameters of structural microfilaments in cells are roughly 6–10 nanometers or $6{-}10 \times 10^{-9}$ m (Fuchs and Cleveland, 1998). Magnified 10^7 times, they would be 6–10 cm or roughly 2–4 inches thick. Mitochondria typically range in length between 0.5 and 10 micrometers or 5×10^{-7} to 10^{-5} m (e.g., Krauss, 2001), converting to 5–100 m, consistent with the size of a tractor-trailer truck in this example.

CHAPTER 2: THE DANCE OF THE ATOMS

Online calculator for molecular speeds and collision rates (HyperPhysics, Department of Physics and Astronomy, Georgia State University): http://hyperphysics.phy-astr.gsu.edu/hbase/kinetic/kintem.html#c3.
IBM Web site for "A Boy and His Atom": http://www.research.ibm.com/articles/madewithatoms.shtml#fbid=yfOjFKDc8us.
Diffusion time calculator on PhysiologyWeb: http://www.physiologyweb.com/calculators/diffusion_time_calculator.html.

CHAPTER 4: CARBON CHAINS

For USEPA information on mercury emissions 1990–2005: http://www.epa.gov/mats/powerplants.html.

For USEPA *information on mercury exposure in women and children*: http:// www.epa.gov/hg/exposure.htm.

For USFDA *information on mercury concentrations in fish*: http://www .fda.gov/Food/FoodborneIllnessContaminants/Metals/ucm115644 .htm.

CHAPTER 5: TEARS FROM THE EARTH

Gluphisia *moth vs. human statistics (after Smedley and Eisner, 1996)*: The male moths weigh about 80 mg and can pump 10–50 ml of water at a sitting. Taking the volume of a moth's emission as 40 ml, the mass of that amount of water is 40,000 mg, which is 500 times larger than 80 mg. Therefore the moths can pump about 500 times their body weight at a sitting. If you weigh 150 pounds, you would therefore pump 75,000 pounds of water (500×150). A gallon of water weighs 8.3 pounds, so this would represent about 9,000 gallons of waste water. A *Gluphisia* moth's body is roughly 1 cm long, and it can pump a jet of waste fluid 40 cm, or 40 body lengths. If you stand 1.8 m tall, 40 of your body lengths would be 72 m, or 236 feet. A *Gluphisia* moth can collect 17 micrograms of sodium at a sitting, and a male moth's total body sodium is normally about 19 micrograms. A moth can therefore collect the equivalent of most of its body sodium supply at a sitting. Adult human body sodium content is roughly 3.4 ounces (Freitas, 1998), so a reasonable estimate of puddled sodium would be 3 ounces.

CHAPTER 6: LIFE, DEATH, AND BREAD FROM THE AIR

User-friendly Web sites on scattering and colors of the sky:
Dietrich Zawischa. "Scattering: The colours of the sky." http://www .itp.uni-hannover.de/~zawischa/ITP/scattering.html; NOAA National Weather Service, JetStream—Online School for Weather. "The Color of Clouds"; http://www.srh.noaa.gov/jetstream/clouds/color.htm. Atmospheric Optics. "Why is the sky blue?" http://www.atoptics.co .uk/atoptics/blsky.htm.

Chapter 8: Limits to Growth

Length of total human DNA strung together: A recent estimate of a total human cell count is 37 trillion (Bianconi et al., 2013). However, red blood cells contain no DNA, and they are abundant in your body. A typical adult carries about 5 liters of blood (you can use the blood-volume calculator at easycalculation.com to adjust for sex, weight, and height), and a typical liter of human blood contains about 5 trillion red blood cells (about 5 million per microliter; Mayo Clinic staff, 2011), which could leave you with about 12 trillion DNA-bearing cells. According to most sources, a human cell contains roughly 2 m of DNA on average, as estimated from the length of two sets of 3 billion human genome base pairs per cell (Annunziato, 2008). Multiplying 12 trillion by 2 m yields a length of 24 billion km (2.4×10^{13} m), or about 15 billion miles. Pluto lies between 4 and 8 billion km away from us, so your combined DNA strand could cover the trip several times.

Sand grains in a swimming pool: If a typical long-course Olympic-size pool measures $50 \, m \times 25 \, m \times 3 \, m$, then its volume is 3.750 m³. If a cube-shaped sand grain has a diameter of 1 mm, then it has a volume of 1 mm³. There are a billion mm³ in a m³. Multiply 3,750 by a billion to get 3.75 trillion such sand grains in such a pool. If you have 7,000 trillion trillion atoms in your body (Freitas, 2008), you could therefore fill about 2 trillion pools with that number of sand grains. If the pools measured 50×25 m apiece, they would cover roughly $(1,250 \, m^2) \times (2 \, trillion) = 25 \times 10^{14} \, m^2$ or $25 \times 10^8 \, km^2$. Earth's surface area is about $5 \times 10^8 \, km^2$, which is about 5 times too small to fit so many pools.

Spending 7,000 trillion trillion dollars: There are 86,400 seconds in a day, and 31.5 million seconds in a year. If you could spend a million dollars per second, you could end up spending 31.5 trillion dollars in a year. Then it would take another 222 trillion years to spend the rest. The current hydrogen-burning, star-producing Stelliferous era of the universe is thought to have an expected duration on the order of 100 trillion years (Adams and Laughlin, 1997).

Amount of potential human carbon in the atmosphere: Trenberth and Smith (2005) put the mass of the atmosphere at 5.1×10^{18} kg, and an atmospheric carbon dioxide concentration of 400 ppm as of 2013 amounts to roughly 2×10^{15} kg of CO_2, with a quarter of that mass due to carbon alone, or roughly 50×10^{13} kg (500 billion metric tons). Freitas (1998) lists an adult's human carbon content as roughly 16 kg, so the carbon content of the atmosphere could supply the carbon budgets of 3.1×10^{13} adult human bodies (31 trillion people).

CHAPTER 9: FLEETING FLESH

Airborne carbon atoms after cremation: A typical adult carries about 8×10^{26} (800 trillion trillion) carbon atoms in his or her body (Freitas, 1998). If they are all converted to CO_2 during cremation and mixed evenly through the atmosphere of the Northern Hemisphere (ca 98 million square miles of land area), there should be (rounding 8×10^{26} atoms to 10^{27} and dividing by 10^8 square miles) on the order of 10^{19} carbon atoms (10 million trillion) over every square mile. There are 27,878,400 square feet in a square mile, so there would be 3.6×10^{11} carbon atoms over a square-foot sector of the Northern Hemisphere, or 360 billion.

References

PROLOGUE

Einstein, A. Quoted in Banesh Hoffmann, *Albert Einstein: Creator and Rebel*. Viking, 1972.

Freitas, R. A. 1998. Nanomedicine. http://www.foresight.org/Nanomedicine/Ch03_1.html.

Hassan, I. 1980. *The Right Promethean Fire*. University of Illinois Press.

Institute of Physics. 2013. Electrons behaving like a particle and a wave: Feynman's double-slit experiment brought to life. *Science-Daily*, 14 Mar. 2013. Web access date, 17 Nov. 2013. http://www.sciencedaily.com/releases/2013/03/130313214031.htm.

Pohl, R., et al. 2010. The size of the proton. *Nature* 466: 213–216.

Slater, J. C. 1964. Atomic radii in crystals. *Journal of Chemical Physics* 41: 3199–3205.

Wilson, E. O. 2002. *The Future of Life*. Knopf Doubleday Publishing Group.

CHAPTER 1: FIRES OF LIFE

Beall, C. M. 2000. Tibetan and Andean patterns of adaptation to high-altitude hypoxia. *Human Biology* 72: 201–228.

Barrett, K. E., S. M. Barman, S. Boitano, and H. Brooks. 2012. *Ganong's Review of Medical Physiology*. McGraw-Hill Medical.

Bell, S. C., M. J. Saunders, J. S. Elborn, and D. J. Shale. 1996. Resting energy expenditure and oxygen cost of breathing in patients with cystic fibrosis. *Thorax* 51: 126–131.

Bender, M. L., M. Battle, and R. F. Keeling. 1998. The O_2 balance of the atmosphere: A tool for studying the fate of fossil-fuel CO_2. *Annual Reviews of Energy and the Environment* 23: 207–223.

Bianconi, E., et al. 2013. An estimation of the number of cells in the human body. *Annals of Human Biology* 40: 463–471.

Boyle, R. 1674. *Suspicions about the Hidden Realities of Air*. In J. F. Fulton, 1932. Robert Boyle and his influence on thought in the seventeenth century. *Isis* 18: 77–102. .

Brennan, D. S. 2013. Keeling curve a mainstay of climate science. *San Diego Union-Tribune*, June 2.

Broecker, W. S. 1996. Breathing easy: Et tu, O_2? *21stC, The World of Research at Columbia University*. Special Issue: Biospheres. http://www.columbia.edu/cu/21stC/issue-2.1/broecker.htm.

———. 1970. Man's oxygen reserve. *Science* 168: 1537–1538.

Children's Hospital Boston. 2012. Injecting life-saving oxygen into a vein. *Science Daily*, June 27, 2102. http://www.sciencedaily.com/releases/2012/06/120627142512.htm.

Daniels, V. G., P. R. Wheater, and H. G. Burkitt. 1979. *Functional Histology: A Text and Colour Atlas*. Churchill Livingstone.

Da Poian, A. T., T. El-Bacha, and M. R. M. P. Luz. 2010. Nutrient utilization in humans: Metabolism pathways. *Nature Education* 3: 11.

deRoetth, A. 1950. Respiration of the cornea. *AMA Archives of Ophthalmology* 44: 666–676.

Dole, M. 1965. The natural history of oxygen. *Journal of General Physiology* 49: 5–27.

Faraday, M. 1861. *A Course of Six Lectures on the Chemical History of a Candle*. Edited by W. Crookes. Griffin, Bohn, & Co.

Freitas, R. A. 1998. Nanomedicine. http://www.foresight.org/Nanomedicine/Ch03_1.html.

Fuchs, E., and D. W. Cleveland. 1998. A structural scaffolding of intermediate filaments in health and disease. *Science* 279: 514–519.

Ganong, W. F. 2012. *Review of Medical Physiology*. McGraw-Hill.

Hansen, J. T., and B. M. Koeppen. 2002. *Netter's Atlas of Human Physiology*. Icon Learning Systems.

Hathaway, D. 2011. The solar interior. Solar Physics: Marshall Space Flight Center. http://solarscience.msfc.nasa.gov/interior.shtml.

Hoffman, J. 2011. Chemical connector. *Nature* 480: 179.

Jackson, S. 1985. *Anatomy & Physiology for Nurses*. Nurses' Aids Series. Bailliere Tindall.

Keeling, R. F. 2008. Recording Earth's vital signs. *Science* 319: 1771–1772.

Keeling, R. F., and S. R. Shertz. 1992. Seasonal and interannual variations in atmospheric oxygen and implications for the global carbon cycle. *Nature* 358: 723–727.

———, and H. E. Garcia. 2002. The change in oceanic O_2 inventory associated with recent global warming. *Proceedings of the National Academy of Sciences* 99: 7848–7853.

Kheir, J. N., et al. 2012. Oxygen gas-filled microparticles provide intravenous oxygen delivery. *Science Translational Medicine*, doi: 10.1126/scitranslmed.3003679.

Kintz, J. 2013. *This Book Title Is Invisible*. Amazon Digital Services, Inc. Quote accessed through *GoodReads*, Dec. 28, 2013. https://www.goodreads.com/quotes/553288-my-next-breath-may-very-well-be-in-your-lungs.

Krauss, S. 2001. Mitochondria: Structure and role in respiration. *Encyclopedia of Life Sciences*. Nature Publishing Group. 6 pp. http://www.med.ufro.cl/clases_apuntes/cs_preclinicas/mg-fisica-medica/sub-modulo-1/Mitochondria.pdf.

Kress, J. P., A. S. Pohlman, J. Alverdy, and J. B. Hall. 1999. The impact of morbid obesity on oxygen cost of breathing (VO_{2resp}) at rest. *American Journal of Respiratory and Critical Care Medicine* 160: 883–886.

Krier, R. 2008. Graphic evidence: Keeling's half-century of CO_2 measurements serves as global warming's longest yardstick. *San Diego Union-Tribune*, March 27.

Lavoisier, A. 1776. *Essays, Physical and Chemical*. Translated from the French by Thomas Henry, with notes and an appendix, 1776. Printed for Joseph Johnson, London, No. 72, Saint Paul's Church-Yard.

Liebig, J. 1843. *Animal Chemistry*. John Owen Publishing, Cambridge, UK.

Lutgens, F. K., and E. J. Tarbuck. 1995. *The Atmosphere*. Prentice-Hall, 6th ed.

Mohr, H., and P. Schopfer. 1995. *Plant Physiology*. Translated from the German by G. Lawlor and D. W. Lawlor. Springer-Verlag. doi:10.1007/978-3-642-97570-7.

Mühfeld, C., E. R., et al. 2010. Is length an appropriate estimator to characterize pulmonary alveolar capillaries? A critical evaluation in the human lung. *Anatomical Record* (Hoboken) 293: 1270–1275.

NASA. Closing the loop: Recycling water and air in space. http://www.nasa.gov/pdf/146558main_RecyclingEDA(final)%204_10_06.pdf.

National Candle Association. 2013. The science of candles. http://www.candles.org/candlescience.html.

Neligan, P. 2002. How much oxygen is in the blood? Critical Care Medicine Tutorials. http://www.ccmtutorials.com/rs/oxygen/page03.htm.

Notter, R. H. 2000. *Lung Surfactants: Basic Science and Clinical Applications*. Marcel Dekker.

Nowak, D. J., R. Hoehn, and D. E. Crane. 2007. Oxygen production by urban trees in the United States. *Arboriculture & Urban Forestry* 33: 220–226.

Palmer, W. 1999. Edison's last breath. *Exquisite Corpse*. http://www.corpse.org/archives/issue_1/palmer.html.

Philips, K. J. H. 1995. *Guide to the Sun*. Cambridge University Press.

Rogoff, G. 2008. Plasma and flames—the burning question. *Coalition for Plasma Science*. http://www.plasmacoalition.org/plasma_writeups/flame.pdf.

Rolfe, D. F., and G. C. Brown. 1997. Cellular energy utilization and molecular origin of standard metabolic rate in mammals. *Physiological Reviews* 77: 731–758.

Satoh, M., and T. Kuroiwa. 1991. Organization of multiple nucleoids and DNA molecules in mitochondria of a human cell. *Experimental Cell Research* 196: 137–140.

Shapley, H. 1967. *Beyond the Observatory*. Charles Scribner's Sons.

Simpson, A. L. E., A. D. Ray, D. R. Pendergast, and C. E. G. Lundgren. 2008. Energy cost of breathing at depth. Abstract of the Undersea & Hyperbaric Medical Society 2008 Annual Scientific Meeting June 26–28, 2008, Salt Lake City Marriott Downtown. http://archive.rubicon-foundation.org/xmlui/handle/123456789/7810.

Slater, J. C. 1964. Atomic radii in crystals. *Journal of Chemical Physics* 41: 3199–3205.

Smusiak, C. 2010. Extreme breath-holding: How it's possible. *Planet Green, HowStuffWorks, Inc.* (February 17, 2010). http://news.discovery.com/human/evolution/breath-holding-human.htm.

Stücker, M., et al. 2002. The cutaneous uptake of atmospheric oxygen contributes significantly to the oxygen supply of human dermis and epidermis. *Journal of Physiology* 538: 985–994.

Trenberth, K. E., and L. Smith. 2005. The mass of the atmosphere: A constraint on global analyses. *Journal of Climate* 18: 864–875.

United States Energy Information Administration. 2013. Frequently asked questions: How much carbon dioxide is produced by burning gasoline and diesel fuel? http://www.eia.gov/tools/faqs/faq.cfm?id=307&t=11.

Walcott, B. The lacrimal gland and its veil of tears. *Physiology* 13: 97–103.

Walker, J. 1978. The physics and chemistry underlying the infinite charm of a candle flame. *Scientific American* 238: 154–162.

WebMD. 2002. The rise of oxygen bars. http://www.webmd.com/balance/features/rise-of-oxygen-bars.

West, J. B. 2002. Highest permanent human habitation. *High Altitude Medical Biology* 3: 401–407.

Zeatoun, L.A., and P. W. Morrison. 1997. Optimizing diamond growth for an atmospheric oxyacetylene torch. *Journal of Materials Research* 12: 1237–1252.

CHAPTER 2: THE DANCE OF THE ATOMS

Blyth, L. 2001. Oxygen isotope analysis and tooth enamel phosphate and its application to archaeology. *Totem: The University of*

Western Ontario Journal of Anthropology 9: Issue 1, Article 2.
http://ir.lib.uwo.ca/totem/vol9/iss1/2.

Bowen, G. J., L. I. Wassenaar, and K. A. Hobson. 2005. Global
application of stable hydrogen and oxygen isotopes to wildlife
forensics. *Oecologia* 143: 337–348.

Bowen, G. J., et al. 2005. Treatment methods for the determination
of δ^2H and $\delta^{18}O$ of hair keratin by continuous-flow isotope-ratio
mass spectrometry. *Rapid Communications in Mass Spectrometry*
19: 2371–2378.

———. 2005. Stable hydrogen and oxygen isotope ratios of bottled
waters of the world. *Rapid Communications in Mass Spectrometry*
19: 3442–3450.

Brown, Robert. 1866. *The Miscellaneous Botanical Works of Robert
Brown*: Vol. 1. Edited by J. J. Bennett. R. Hardwicke.

Carus, Titus Lucretius. 1851. *The Nature of Things*. Literally translated
into English prose, by the Rev. John Selby Watson, to which is
adjoined the poetical version of John Mason Good. H. G. Bohn.

———. 2006. Stable isotopes in elephant hair documents migration
patterns and diet changes. *Proceedings of the National Academy of
Sciences* 103: 371–373.

Cerling, T. E., et al. 2009. History of animals using isotope records
(HAIR): A 6-year dietary history of one family of African
elephants. *Proceedings of the National Academy of Sciences*
106: 8093–8100.

Chamberlain, C. P., et al. 1997. The use of isotope tracers for identifying
populations of migratory birds. *Oecologia* 109: 132–141.

Chesson, L. A., et al. 2010. Links between purchase location and
the stable isotope ratios of bottled water, soda, and beer in
the United States. *Journal of Agricultural and Food Chemistry*
58: 7311–7316.

Cho, R. 2011. From wastewater to drinking water. Posted April 4,
2011, on *State of the Planet* blog. Earth Institute, Columbia
University. http://blogs.ei.columbia.edu/2011/04/04/from
-wastewater-to-drinking-water/.

Ehleringer, J. R., et al. 2008. Hydrogen and oxygen isotope ratios in human hair are related to geography. *Proceedings of the National Academy of Sciences* 105: 2788–2793.

Einstein, A. 1905. Investigations on the theory of the Brownian movement. Originally published in 1905 in *Annalen der Physik* 17, p. 594. Translated from the German by A. D. Cowper, edited with notes by R. Fürth. Dover Publications (1956 ed.). http://www.pitt.edu/~jdnorton/lectures/Rotman_Summer_School_2013/Einstein_1905_docs/Einstein_Brownian_English.pdf.

Eiseley, L. 1957. *The Immense Journey: An Imaginative Naturalist Explores the Mysteries of Man and Nature*. Vintage Books.

Fisher, D. C. 2009. Paleobiology and extinction of proboscideans in the Great Lakes region of North America. In *American Megafaunal Extinctions at the End of the Pleistocene*. Edited by G. Haynes, 55–75. Springer.

———, S. G. Beld, and A. N. Rountrey. 2008. Tusk record of the North Java Mastodon. In Mastodon paleobiology, taphonomy, and paleoenvironment in the late Pleistocene of New York State: Studies on the Hyde Park, Chemung, and North Java sites. Edited by W. D. Allmon and P. L. Nester. *Palaeontographica Americana* 61: 417–463.

Ford, B. J. 1992. Brownian movement in *Clarkia* pollen: A reprise of the first observations. *The Microscope* 40: 235–241.

Fricke, H. C., J. R. O'Neill, and N. Lynnerup. 1995. Oxygen isotope composition of human tooth enamel from medieval Greenland: Linking climate and society. *Geology* 23: 869–872.

Friedman, I., A. C. Redfield, B. Schoen, and J. Harris. 1964. The variation of the deuterium content of natural waters in the hydrologic cycle. *Reviews of Geophysics* 2: 177–223.

Fumagalli, M., J. M. O'Meara, and J. X. Prochaska. 2011. Detection of pristine gas two billion years after the Big Bang. *Science* 334: 1245–1249.

Greenblatt, S. 2012. *The Swerve: How the World Became Modern*. W. W. Norton.

Griffith, C. A., et al. 2012. Possible tropical lakes on Titan from observations of dark terrain. *Nature* 486: 237–239.

Hobson, K. A. 1999. Tracing origins and migration of wildlife using stable isotopes: A review. *Oecologia* 120: 314–326.

———, R. Barnett-Johnson, and T. Cerling. 2010. Using isoscapes to track animal migration. In *Isoscapes: Understanding Movement, Pattern, and Process on Earth through Isotope Mapping*. Edited by J. B. West et al., chap. 13. Springer Science + Business Media B.V.

Hosokawa, T., K. Omukai, N. Yoshida, and H. W. Yorke. 2011. Protostellar feedback halts the growth of the first stars in the universe. *Science* 334: 1250–1253.

Kennedy, C. D., G. J. Bowen, and J. R. Ehleringer. 2011. Temporal variations of oxygen isotope ratios ($d^{18}O$) in drinking water: Implications for specifying location of origin with human scalp hair. *Forensic Science International* 208: 156–166.

Knudson, K. J. 2009. Oxygen isotope analysis in a land of environmental extremes: The complexities of isotopic work in the Andes. *International Journal of Osteoarchaeology* 19: 171–191.

Koch, P. L., D. C. Fisher, and D. Dettman. 1989. Oxygen isotope variation in the tusks of extinct proboscideans: A measure of season of death and seasonality. *Geology* 17: 515–519.

Lahnsteiner, J., and G. Lempert. 2005. Water management in Windhoek/Namibia. *Proceedings of the IWA Conference, Wastewater Reclamation & Reuse for Sustainability*, November 8–11, 2005, Jeju, Korea.

Lappert, M. F., and J. N. Murrell. 2003. John Dalton, the man and his legacy: The bicentenary of his Atomic Theory. *Dalton Transactions* 3811–3820. doi: 10.1039/b307622a.

Law, I. B. 2003. Advanced reuse: From Windhoek to Singapore and beyond. *Water* 30: 31–36.

Lee, J. Y., J. W. Choi, and H. Kim. 2007. Determination of hand surface area by sex and body shape using alginate. *Journal of Physiological Anthropology* 26: 475–483.

Longinelli, A. 1984. Oxygen isotopes in mammal bone phosphate: A new tool for paleohydrological and paleoclimatological research? *Geochimica et Cosmochimica Acta* 48: 385–390.

Loschmidt, J. 1865. On the size of air molecules. *Proceedings of the National Academy of Science of Vienna* 52: 395–413.

Paige, D. 2009. Diviner Commissioning Observations. Web site of DIVINER Lunar Radiometer Experiment. Posted September 17, 2009. http://www.diviner.ucla.edu/blog/?p=123.

Podlesak, D. W., et al. 2012. d^{2}H and d^{18}O of human body water: A GIS model to distinguish residents from non-residents in the contiguous USA. *Isotopes in Environmental and Health Studies* 48:259–279.

Rumi, J. M. Quoted in Clark, R. Y., 2001 *Stranger Gods: Salman Rushdie's Other Worlds.* McGill-Queen's University Press.

Sharp, Z. D., V. Atudorei, H. O. Panarelllo, J. Fernandez, and C. Douthitt. 2003. Hydrogen isotope systematics of hair: Archaeological and forensic applications." *Journal of Archaeological Science* 30: 1709–1716.

Silverberg, J. L., M. Bierbaum, J. P. Sethna, and I. Cohen. 2013. Collective motion of moshers at heavy metal concerts. *arXiv*:1302.1886v1. doi:10.1103/PhysRevLett.110.228701.

Smith, K., and D. Fisher. 2011. Sexual dimorphism of structures showing indeterminate growth: Tusks of American mastodons (*Mammut americanum*). *Paleobiology* 37: 175–194.

Tempaku, A. 2005. Random Brownian motion regulates the quantity of Human Immunodeficiency Virus Type-1 (HIV-1) attachment and infection to target cell. *Journal of Health Sciences* 51: 237–241.

Tyson, P. J. 2001. Molecular Collisions. The Kinetic Atmosphere. http://hyperphysics.phy-astr.gsu.edu/hbase/kinetic/kintem.html#c3.

Wolf, B. O., and C. Martinez del Rio. 2000. Use of saguaro fruit by white-winged doves: Isotopic evidence of a tight ecological association. *Oecologia* 124: 536–543.

Chapter 3: Blood Iron

Aguirre, J. D., et al. 2013. A manganese-rich environment supports superoxide dismutase activity in a Lyme Disease pathogen. *Borrelia burgdorferi. Journal of Biological Chemistry* 288: 8468–8478.

Albretsen, J. 2006. The toxicity of iron, an essential nutrient. *Veterinary Medicine* 101: 82–90.

Aust, S. D. 1995. Ferritin as a source of iron and protection from iron-induced toxicities. *Toxicology Letters* 82: 941–944.

Barlow, M. J., et al. 2013. Detection of a noble gas molecular ion, $^{36}ArH^+$, in the Crab Nebula. *Science* 342: 1343–1345.

Battacharjee, Y. 2012. How do stars explode? *Science* 336: 1094–1095.

Berlim, M. T. 2001. Evolutionary approach to medicine. *Southern Medical Journal* 94: 26–32.

Ben-Ami, Y., I. Koren, Y. Rudich, S. T. Martin, and M. O. Andreae. 2010. Transport of North African dust from the Bodélé depression to the Amazon basin: A case study. *Atmospheric Chemistry and Physics Discussions* 10: 4345–4372.

Breen, A., and D. McCarthy. 1995. A re-evaluation of the Eastern and Western records of the supernova of 1054. *Vistas in Astronomy* 39: 363–379.

Cawein, M., and E. J. Lappat. 1964. Hereditary ethemoglobinemia. In *Hemoglobin, Its Precursors and Metabolites* Edited by F. William Sunderman. J. B. Lippincott Co.

———, C. H. Behlen, E. J. Lappat, and J. E. Cohn. 1964. Hereditary diaphorase deficiency and methemoglobinemia. *Archives of Internal Medicine* 113: 578–585.

Centers for Disease Control and Prevention. 1993. Toddler deaths resulting from ingestion of iron supplements—Los Angeles, 1992–1993. *Morbidity and Mortality Weekly Report* 42: 111–113.

Cescutti, G., F. Matteucci, E. Caffau, and P. François. 2012. Chemical evolution of the Milky Way: The origin of phosphorus (Research Note). *Astronomy & Astrophysics* 540, 4 pp. doi: 10.1051/0004-6361/201118188.

Clarke, D. A. 2012. A truly embryonic star. *Nature* 492: 52–53.

Daniel, J. 2012. *Of Earth*. Lost Horse Press.

Dohnanyi, J. S. 1972. Interplanetary objects in review: Statistics of their masses and dynamics. *Icarus* 17: 1–48.

Edmonds, P. 2013. The remarkable properties of neutron stars.

(Blog post on Chandra X-Ray Observatory Web site.)
http://www.chandra.harvard.edu/blog/node/432.

FAO/WHO. 2002. Iron. In *Human Vitamin and Mineral Requirements*,
chap. 13. Report of a joint FAO/WHO expert consultation,
Bangkok, Thailand. http://www.fao.org/docrep/004/y2809e/
y2809e0j.htm.

Fleming, R. E., and B. R. Bacon. 2005. Orchestration of iron
homeostasis. *New England Journal of Medicine* 352: 1741–1744.

Fownes, G. 1863. *A Manual of Elementary Chemistry, Theoretical and
Practical.* J. Churchill and Sons.

Gail, H.-P., and E. Sedlmayr. 1999. Mineral formation in stellar
winds. *Astronomy and Astrophysics* 347: 594–616.

Hamblin, T. J. 1981. Fake! *British Medical Journal* 283: 1671–1674.

Heger, A. 2013. Going supernova. *Nature* 494: 46–47.

Helmer, M., J. M. C. Plane, J. Qian, and C. S. Gardner. 1998.
A model of meteoric iron in the upper atmosphere. *Journal
of Geophysical Research* 103: 10,913–10,925.

Holmes, M. A., et al. 2005. Siderocalin (Lcn 2) also binds
carboxymycobactins, potentially defending against mycobacterial
infections through iron sequestration. *Structure* 13: 29–41.

Hosokawa, T., K. Omukai, N. Yoshida, and H. W. Yorke. 2011.
Protostellar feedback halts the growth of the first stars in the
universe. *Science* 334: 1250–1253.

Hunt, J. R., C. A. Zito, and L. K. Johnson. 2009. Body iron excretion
by healthy men and women. *American Journal of Clinical Nutrition*
89: 1792–1798.

Höfner, S. 2012. Fresh light on stardust. *Nature* 484: 172–173.

Isaacson, W. 2008. *Einstein: His Life and Universe.* Simon & Schuster.

Kacprzak, G. G. 2011. The pristine universe. *Science* 334: 1216–1217.

Kirschvink, J. L., M. M. Walker, and C. E. Diebel. 2001. Magnetite-based
magnetoreception. *Current Opinions in Neurobiology* 11: 462–467.

Koo, B.-C., et al. 2013. Phosphorus in the young supernova remnant
Cassiopeia A. *Science* 342: 1346–1348.

Koren, I., et al. 2006. The Bodélé depression: A single spot in the
Sahara that provides most of the mineral dust to the Amazon.

Environmental Research Letters 1: 014005. doi 10.1088/1748-9326/1/1/014005.

Lopes, T. J. S., et al. 2010. Systems analysis of iron metabolism: The network of iron pools and fluxes. *BMC Systems Biology* 4: 112–130.

MacAlpine, G. M., and T. J. Satterfield. 2008. The Crab Nebula's composition and precursor star mass. *Astronomical Journal* 136: 2152. doi:10.1088/0004-6256/136/5/2152.

McDonald, I., et al. 2010. Rusty old stars: A source of the missing interstellar iron? *Astrophysical Journal Letters* 717: L92–L97.

Morris, E. R. 1983. An overview of current information on bioavailability of dietary iron to humans. *Federation for American Societies for Experimental Biology* 42: 1716–1720.

Motizuki, Y., et al. 2009. An Antarctic ice core recording both supernovae and solar cycles. *arXive:0902.3446*.

Murdin, P., and L. Murdin. 1985. *Supernovae.* Cambridge University Press.

Murray, R. W., M. Leinen, and C. W. Knowlton. 2012. Links between iron input and opal deposition in the Pleistocene equatorial Pacific Ocean. *Nature Geoscience* 5: 270–274.

Music Trades. 2010. Ulbrich Steel—an unsung hero. Music Trades Corp article republished by *HighBeam Research*, November 10, 2013. http://www.highbeam.com/doc/1G1-230156328.html.

Needham, J. 1986. *Science and Civilization in China.* Vol. 5, *Chemistry and Chemical Technology.* Part 7, *Military Technology: The Gunpowder Epic.* Caves Books, Ltd.

Norris, B. R. M., et al. 2012. A close halo of large transparent grains around extreme red stars. *Nature* 484: 220–222.

Nugent, P. E., et al. 2011. Supernova SN 2011fe from an exploding carbon-oxygen white dwarf star. *Nature* 480: 344–346.

Pankenier, D. W. 2006. Notes on translations of the East Asian records relating to the supernova of AD 1054. *Journal of Astronomical History and Heritage* 9: 77–82.

Perry, R. D., P. B. Balbo, H. A. Jones, J. D. Fetherston, and E. Demoll. 1999. Yersiniabactin from Yersinia pestis: Biochemical characterization of the siderophore and its role in iron transport and regulation. *Microbiology* 145: 1181–1190.

Pocaro, V. F., and A. Martocchia. 2006. Supernovae astrophysics from Middle Age documents. *Proceedings International Astronomical Union*. Symposium No. 235, 2005. doi:10.1017/S1743921306008416.

Popova, O. P., et al. 2013. Chelyabinsk airburst, damage assessment, meteorite recovery, and characterization. *Science* 342: 1069–1073.

Powell, D. 2011. Sparing the rare earths. *Science News* 180: 18–21.

Prufer, O. H. 1961. Prehistoric Hopewell meteorite collecting: Context and implications. *Ohio Journal of Science* 61: 341–352.

Reilly, C. 2004. *The Nutritional Trace Metals*. Blackwell Publishing, Oxford, UK.

Richier, S., et al. 2012. Abundances of iron-binding photosynthetic and nitrogen-fixing proteins of *Trichodesmium* both in culture and *in situ* from the North Atlantic. PLOS ONE 7:e35571.

Risbo, T., H. B. Clausen, and K. L. Rasmussen. 1981. Supernovae and nitrate in the Greenland Ice Sheet. *Nature* 294: 637–639.

Roebroeks, W., et al. 2012. Use of red ochre by early Neanderthals. *Proceedings of the National Academy of Sciences* 109: 1889–1894.

Royal Astronomical Society (RAS). 2012. Measuring the cosmic dust swept up by Earth. *Science Daily*, 29 March, 2012. http://www.sciencedaily.com/releases/2012/03/120329225140.htm.

Sardi, B. 2000. Iron: Too much of a good thing. *Nutrition Science News* (June 2000). http://www.chiro.org/nutrition/FULL/Iron_Too_Much_of_a_Good_Thing.shtml.

Stachel, J. 2000. Introduction, to *Einstein: The Formative Years, 1879–1909*. Edited by D. Howard and J. Stachel, pp. 1–22. Birkhäuser.

Stapput, K., et al. 2010. Magnetoreception of directional information in birds requires nondegraded vision. *Current Biology* 20: 1–4.

Sutton, M. 2010. Spinach, iron and Popeye: Ironic lessons from biochemistry and history on the importance of healthy eating, healthy scepticism and adequate citation. *Internet Journal of*

Criminology (Primary Research Paper series). http://www
.internetjournalofcriminology.com/Sutton_Spinach_Iron_and
_Popeye_March_2010.pdf.

———. 2012. The spinach, Popeye, iron, decimal error myth is
finally busted. http://www.bestthinking.com/articles/science/
chemistry/biochemistry/the-spinach-popeye-iron-decimal-error
-myth-is-finally-busted.

Thielk, D. 2000. The birth of the elements. *ChemMatters* (October,
2000), pp. 4–6.

Treiber, C. D., et al. 2012. Clusters of iron-rich cells in the upper
beak of pigeons are macrophages not magnetosensitive neurons.
Nature 484: 367–370.

Trost, C. 1982. The blue people of Troublesome Creek. *Science 82*.
http://www.nclark.net/BluePeopleofTroublesomeCreek.html.

Tuck, C. A., and R. L. Virta. 2011. Iron ore. In United States Geological
Survey, 2011. *Minerals Yearbook*. http://minerals.usgs.gov/minerals/
pubs/commodity/iron_ore/myb1-2011-feore.pdf.

Washington, R., et al. 2009. Dust as a tipping element: The Bodélé
depression, Chad. *Proceedings of the National Academy of Sciences*
106: 20,564–20,571.

Winklhofer, M. 2012. An avian magnetometer. *Nature* 336: 991–992.

Woosley, S. E., A. Heger, and T. A. Weaver. 2002. The evolution and
explosion of massive stars. *Reviews of Modern Physics* 74: 1015–1071.

Yanatori, I., et al.. 2010. Heme and non-heme iron transporters in
non-polarized and polarized cells. *BMC Cell Biology* 11: 39.

Zhou, G., et al. 2007. Fate of blood meal iron in mosquitoes. *Journal
of Insect Physiology* 53: 69–1178.

Zinner, E. 2003. An isotopic view of the early solar system.
Science 300: 265–267.

CHAPTER 4: CARBON CHAINS

Adams, M. Quoted in V. Mougios, *Exercise Biochemistry*, 2006.
Human Kinetics Publishers, Champaign.

Affek, H. P., and J. M. Eiler. 2006. Abundance of mass 47 CO_2 in urban air, car exhaust, and human breath. *Geochimica et Cosmochimica Acta* 70: 1–12.

Arneborg, J., et al. 1999. Change of diet of the Greenland Vikings determined from stable carbon isotope analysis and ^{14}C dating of their bones. *Radiocarbon* 41: 157–168.

Briber, B. M., et al. 2013. Variations in atmospheric CO2 mixing ratios across a Boston, MA, urban to rural gradient. *Land* 2: 304–327.

Bush, S. E., D. E. Pataki, and J. R. Ehleringer. 2007. Sources of variation in $\delta^{13}C$ of fossil fuel emissions in Salt Lake City USA. *Applied Geochemistry* 22: 715–723.

Centers for Disease Control and Prevention. 2004 Blood mercury levels in young children and child-bearing-aged women—United States, 1999–2002. *Morbidity and Mortality Weekly Report* 53: 1018–1020.

Chesson, L. A., J. R. Ehleringer, and T. E. Cerling. 2012. Light-element isotopes (H, C, N, and O) as tracers of human diet: A case study on fast food meals, pp. 707–723. In *Handbook of Environmental Isotope Geochemistry: Advances in Isotope Geochemistry.* Edited by M. Baskaran, chap. 33. Springer-Verlag.

Coplen, T. B. 1994 Reporting of stable hydrogen, carbon and oxygen isotopic abundances. *Pure and Applied Chemistry* 66: 273–276.

Day, P. 1999. *The Philosopher's Tree: A Selection of Michael Faraday's Writings.* Institute of Physics Publishing, Bristol, UK.

Doyle, P., and M. R. Bennett. 1995. Belemnites in biostratigraphy. *Palaeontology* 38: 815–829.

Djuricin, S., X. Xu, and D. E. Pataki. 2012. The radiocarbon composition of tree rings as a tracer of fossil fuel emissions in the Los Angeles basin: 1980–2008. *Journal of Geophysical Research Atmospheres* 117, D12,27, doi: 10.1029/2011JD017284.

Falkowski, P., et al. 2000. The global carbon cycle: A test of our knowledge of Earth as a system. *Science* 290: 291–296.

Freitas, R. A. 1998. Nanomedicine. http://www.foresight.org/Nanomedicine/Ch03_1.html.

Gerlach, T. 2011. Volcanic versus anthropogenic carbon dioxide. *EOS* 92: 201–22.

Goulden, M. L., et al. 1996. Exchange of carbon dioxide by a deciduous forest: Response to interannual climate variability. *Science* 271: 1576–1578.

Jim, S., S. H. Ambrose, and R. P. Evershed. 2004. Stable carbon isotopic evidence for differences in the dietary origin of bone cholesterol, collagen, and apatite: Implications for their use in paleodietary reconstruction. *Geochimica et Cosmochimica Acta* 68: 61–72.

Kèlomé, N. C., et al. 2006. C4 plant isotopic composition (delta 13C) evidence for urban CO2 pollution in the city of Cotonou, Benin (West Africa). *Science of the Total Environment* 366: 439–447.

Kennedy, C., et al. 2009. Greenhouse gas emissions from global cities. *Environmental Science and Technology* 49: 7297–7302.

Koerner, B., and J. Klopatek. 2002. Anthropogenic and natural CO_2 emission sources in an arid urban environment. *Environmental Pollution* 116, Supplement 1: S45–S51.

Levi, P. 1984. *The Periodic Table*. Translated from the Italian by R. Rosenthal. Schocken Books.

Mailer, N. 1968. *The Idol and the Octopus*. Dell Bantam.

Martinelli, L. A., et al. 2011. Worldwide stable carbon and nitrogen isotopes of *Big Mac* patties: An example of a truly "glocal" food. *Food Chemistry* 127: 1712–1718.

McKain, K., et al. 2012. Assessment of ground-based atmospheric observations for verification of greenhouse gas emissions from urban areas. *Proceedings of the National Academy of Science* 109: 8423–8428.

Miller, J. B., et al. 2012. Linking emissions of fossil fuel CO_2 and other anthropogenic race gases using atmospheric $^{14}CO_2$. *Journal of Geophysical Research* 117, D08302, doi: 10.1029/2011JD017048.

Nardoto, G. B., et al. 2006. Geographical patterns of human diet derived from stable-isotope analysis of fingernails. *American Journal of Physical Anthropology* 131: 137–146.

Newman, S., et al. 2008. Changes in mixing ratio and isotopic composition of CO_2 in urban air from the Los Angeles basin, California, between 1972 and 2003. *Journal of Geophysical Research* 113, D23304, doi: 10.1029/2008JD009999.

Newman, S., et al. 2012. Diurnal tracking of anthropogenic CO_2 emissions in the Los Angeles basin megacity during spring, 2010. *Atmospheric Chemistry and Physics Discussions* 12: 5771–5801.

Newsome, S. D., et al.. Stable isotopes evaluate exploitation of anthropogenic foods by the endangered San Joaquin kit fox (*Vulpes macrotis mutica*). *Journal of Mammalogy* 91: 1313–1321.

Pataki, D. E., J. R. Ehleringer, and J. M. Zobitz. 2006. High resolution monitoring of urban carbon dioxide sources. *Geophysical Research Letters* 33, L03818, doi:10.1029/2005GL024822.

———, et al. 2006. Urban ecosystems and the North American carbon cycle. *Global Change Biology* 12: 2092–2101.

———, D. R. Bowling, and J. R. Ehleringer. 2003. Seasonal cycle of carbon dioxide and its isotopic composition in an urban atmosphere: Anthropogenic and biogenic effects. *Journal of Geophysical Research—Atmospheres* 108 (D23)4735, doi:10.1029/2003JD003865.

Pawelczyk, S., and A. Pazdur. 2004. Carbon isotopic composition of tree rings as a tool for biomonitoring CO_2 level. *Radiocarbon* 46: 701–719.

Riebeek, C. 2011. The carbon cycle. NASA Earth Observatory. http://earthobservatory.nasa.gov/Features/CarbonCycle/.

Riley, W. J., et al. 2008. Where do fossil fuel carbon dioxide emissions from California go? An analysis based on radiocarbon observations and an atmospheric transport model. *Journal of Geophysical Research Biogeosciences* 113, G04002, doi:10.1029/2007JG000625.

Schuur, E. A. G., and B. Abbott. 2011. High risk of permafrost thaw. *Nature* 480: 32–33.

Sherman, L. S., et al. 2012. Investigation of local mercury deposition from a coal-fired power plant using mercury isotopes. *Environmental Science and Technology* 46: 382–390.

Strong, C., et al. 2011. Urban carbon dioxide cycles within the Salt Lake Valley: A multiple box model validated by observations. *Journal of Geophysical Research—Atmospheres* 116, D15307, doi:10.1029/2011JD015693.

United States Energy Information Administration. 2013. U.S.

Energy-Related Carbon Dioxide Emissions, 2012. http://www.eia
.gov/environment/emissions/carbon/pdf/2012_co2analysis.pdf.

Valenzuela, L. O., et al. 2012. Dietary heterogeneity among western
industrialized countries reflected in the stable isotope ratios of
human hair. PLOS ONE 7: e34234. doi: 10.1371/journal.
pone.0034234.

CHAPTER 5: TEARS FROM THE EARTH

Abrahams, P. 2003. Human geophagy: A review of its distribution,
causes, and·implications. In *Geology and Health: Closing the Gap*.
Edited by H. Catherine, W. Skinner, A. R. Berger, 31–36. Oxford
University Press, NY.

Ayotte, J. B., K. L. Parker, J. M. Arocena, and M. P. Gillingham. 2006.
Chemical composition of lick soils: Functions of soil ingestion by
four ungulate species. *Journal of Mammalogy* 87: 878–888.

Bänziger, H. 1992. Remarkable new cases of moths drinking
human tears in Thailand (Lepidoptera: Thyatridae, Sphingidae,
Notodontidae). *Natural History Bulleting of the Siam Society*
40: 91–102.

———. 1990. Moths with a taste for tears. *New Scientist* 11: 48–51.

Bates, G. P., and V. S. Miller. 2008. Sweat rate and sodium loss during
work in the heat. *Journal of Occupational Medicine and Toxicology*
3:4, doi: 10.1186/1745-6673-3-4.

Callahan, K. L. 2000. Pica, geophagy, and rock art: Ingestion of rock
powder and clay by humans and its implications for the production
of some rock art on a global basis. Paper read at Society of
American Archivists Conference, Philadelphia. April 8, 2000.
http://www.tc.umn.edu/~call0031/pica.html.

Chaudhari, N., and S. D. Roper. The cell biology of taste. *Journal of
Cell Biology* 190: 285–296.

Clarke, S., and R. McHugh. 2009. Jury rules against radio station after
water-drinking contest kills Calif. mom. ABC News, Good
Morning America. http://abcnews.go.com/GMA/jury-rules-radio
-station-jennifer-strange-water-drinking/story?id=8970712.

Cunningham, J. H., G. Milligan, and L. Trevisan. 2004. Minerals in Australian fruits and vegetables—a comparison of levels between the 1980s and 2000. Food Standards Australia New Zealand. http://www.foodstandards.gov.au/publications/documents/minerals_report.pdf.

Diamond, J. 1998. Eat dirt! *Discover*, February, 1998. pp. 70–75.

Dinesen, I. 1934. (Karen Blixen) *Seven Gothic Tales*. Harrison Smith & Robert Haas.

Fennelly, B. A. 2010. Beth Ann Fennelly digs into geophagy. *Oxford American*, Issue 68. http://www.oxfordamerican.org/articles/2010/mar/09/wide-world-eating-dirt/.

Frassetto, L. A., R. C. Morris, D. E. Sellmeyer, and A. Sebastian. 2008. Adverse effects of sodium chloride on bone in the aging human population resulting from habitual consumption of typical American diets. *The Journal of Nutrition* 138: 419S–422S.

Freitas, R. A. 1998. Nanomedicine. http://www.foresight.org/Nanomedicine/Ch03_1.html.

Harmer, P. M., and E. J. Benne. 1945. Sodium as a crop nutrient. *Soil Science* 60: 137–149.

Hartline, D. K., and D. R. Colman. 2007. Rapid conduction and the evolution of giant axons and myelinated fibers. *Current Biology* 17: R29–R35.

Hilgartner, R., et al. 2007. Malagasy birds as hosts for eye-frequenting moths. *Biology Letters* 3: 117–120.

Janáček, K., and K. Siegler. 2000. Osmosis: Membranes impermeable and permeable for solutes, mechanism of osmosis across porous membranes. *Physiological Research* 49: 191–195.

Kalat, J. W. 1998. *Biological Psychology*, 6th ed. Wadsworth.

Kendrick, M. A., M. Scambelluri, M. Honda, and D. Phillips. 2011. High abundances of noble gas and chlorine delivered to the mantle by serpentinite subduction. *Nature Geoscience* 4: 807–812.

Kreulen, D. A. 1985. Lick use by large herbivores: A review of benefits and banes of soil consumption. *Mammal Review* 15: 107–123.

Kuroda, P. K., and E. B. Sandell. 1963. Chlorine in igneous rocks:

Some aspects of the geochemistry of chloride. *Bulletin of the Geological Society of America* 64: 879–896.

Lindemann, B. 2001. Receptors and transduction in taste. *Nature* 413: 219–225.

Maughan, R. J., and J. B. Leiper. 1995. Sodium intake and post-exercise rehydration in man. *European Journal of Applied Physiology* 71: 311–319.

McCance, R. A. 1936. Experimental sodium chloride deficiency in man. *Proceedings of the Royal Society of London* B 119: 245–268. doi:10.1098/rspb.1936.0009.

Miller, I. J., and F. E. Reedy, Jr. 1990. Variations in human taste bud density and taste intensity perception. *Physiology & Behavior* 47: 1213–1219.

Mills, A., and A. Milewski. 2007. Geophagy and nutrient supplementation in the Ngorongoro Conservation Area, Tanzania, with particular reference to selenium, cobalt, and molybdenum. *Journal of Zoology* 271: 110–118.

Molleman, F., et al. 2005. Is male puddling behaviour of tropical butterflies targeted at sodium for nuptial gifts or activity? *Biological Journal of the Linnean Society* 86: 345–361.

Oka, Y., et al. 2013. High salt recruits aversive taste pathways. *Nature* 494: 472–475.

Pardo, J. M., and F. J. Quintero. 2002. Plants and sodium ions: Keeping company with the enemy. *Genome Biology* 3, reviews1017.1-reviews117.4.

Richardson, J. A. 2000. Permethrin spot-on toxicoses in cats. *Journal of Veterinary Emergency and Critical Care* (April-June, 2000). http://aspcapro.org/sites/pro/files/d-veccs_april00_0.pdf.

Sculley, C. E., and C. L. Boggs. 1996. Mating systems and sexual division of foraging effort affect puddling behavior by butterflies. *Ecological Entomology* 21: 193–197.

Sedov, S. A., et al. 2011. Lysis of *Escherichia coli* cells by lysozyme: Discrimination between adsorption and enzyme action. *Colloids and Surfaces B: Biointerfaces* 88: 131–133.

Selleck, B. 2010. Stratigraphy of the northern Appalachian basin,

Mohawk valley, central New York State. Unpublished report.
http://www.colgate.edu/portaldata/imagegallerywww/089719f7
-d57d-4b54-8523-5331fb2990d9/ImageGallery/paleozoic
-stratigraphy-field-guide-mohawk-valley.pdf.

Shiklomanov, I. 1993. World fresh water resources. In *Water in Crisis: A Guide to the World's Fresh Water Resources*. Edited by P. H. Gleick, 13–24. Oxford University Press, NY.

Smedley, S. R., and T. Eisner. 1996. Sodium: A male moth's gift to its offspring. *Proceedings of the National Academy of Sciences* 93: 809–813.

Steinbach, H. B., and S. Spiegelman. 1943. The sodium and potassium balance in squid nerve axoplasm. *Journal of Cellular and Comparative Physiology* 22: 187–96.

Stetson, D. S., J. W. Albers, B. A. Silverstein, and R. A. Wolfe. 1992. Effects of age, sex, and anthropometric factors on nerve conduction measures. *Muscle & Nerve* 15: 1095–1104.

Stoddart, D. M. 1990. *The Scented Ape: The Biology and Culture of Human Odour*. Cambridge University Press, UK.

Subbarao, G. V., O. Ito, W. L. Berry, and R. M. Wheeler. 2003. Sodium—A functional plant nutrient. *Critical Reviews in Plant Sciences* 22: 391–416.

Swenson, H. 1983. Why is the ocean salty? *U.S. Geological Survey Publication*. http://www.palomar.edu/oceanography/salty_ocean .htm.

Vermeer, D. E., and D. A. Frate. 1975. Geophagy in a Mississippi county. *Annals of the Association of American Geographers* 65: 414–416.

Voigt, C. C., et al. 2008. Nutrition or detoxification: Why bats visit mineral licks of the Amazonian rainforest. PLOS ONE 3: e2011. doi: 10.1371/journal.pone.0002011.

Walcott, B. The lacrimal gland and its veil of tears. *Physiology* 13: 97–103.

Willis, W. D., and R. G. Grossman. 1981. *Medical Neurobiology*. Mosby.

World Health Organization. 1993. Chloride in drinking water. Background document for development *Guidelines for*

Drinking-water Quality. WHO/SDE/WSH/03.04/03. http://www
.who.int/water_sanitation_health/dwq/chloride.pdf.

Wortmann, U. G., and A. Paytan. 2012. Rapid variability of seawater
chemistry over the past 130 million years. *Science* 337: 334–337.

Yotsu-Yamashita, M., et al. Variability of tetrodotoxin and of its
analogues in the red-spotted newt, Notophthalmus viridescens
(Amphibia: Urodela: Salamandridae). *Toxicon* 59: 247–260.

Young, J. Z. 1996. In *The History of Neuroscience in Autobiography*.
Volume 1. Edited by L. R. Squire, 554–586. Oxford University
Press, NY.

CHAPTER 6: LIFE, DEATH, AND BREAD FROM THE AIR

Belnap, J. 2012. Unexpected uptake. *Nature Geoscience* 5: 443–444.

Bohren, C. F., and A. B. Fraser. 1985. Colors of the sky. *The Physics
Teacher* (May issue), pp. 267–272.

Bowlby, C. 2011. Fritz Haber: Jewish chemist whose work led to
Zyklon B. BBC News Europe (April 11, 2011). http://www.bbc.co
.uk/news/world-13015210.

Brans, Y. W., and P. Ortega. 1978. Perinatal nitrogen accretion in
muscles and fingernails. *Pediatric Research* 12: 849–852.

Cabana, G., and J. B. Rasmussen. 1996. Comparison of aquatic food
chains using nitrogen isotopes. *Proceedings of the National Academy
of Sciences* 93: 10,844–10,847.

Canfield, D. E., A. N. Glazer, and P. G. Falkowski. 2010. The evolution and future of Earth's nitrogen cycle. *Science* 330: 192–196.

Charles, D. 2002. The tragedy of Fritz Haber. National Public Radio
(Morning Edition, July 11, 2002). http://www.npr.org/programs/
morning/features/2002/jul/fritzhaber/.

Christensen, J. R., et al. 2005. Persistent organic pollutants in British
Columbia grizzly bears: Consequences of divergent diets.
Environmental Science and Technology 39: 6952–6960.

Cordifi, S. F. 2013. The colors of sunset and twilight. NOAA Storm
Prediction Center Web site http://www.spc.noaa.gov/publications/
corfidi/sunset/.

Crookes, William. 1899. *The Wheat Problem: Based on Remarks Made in the Presidential Address to the British Association at Bristol in 1898.* Longmans, Green, & Co.

Darimont, C. T., and T. E. Reimchen. 2002. Intra-hair stable isotope analysis implies seasonal shift to salmon in gray wolf diet. *Canadian Journal of Zoology* 80: 1638–1642.

Drake, D. C., and R. J. Naiman. 2007. Reconstruction of Pacific salmon abundance from riparian tree-ring growth. *Ecological Applications* 17: 1523–1542.

Einstein, A. 1910. Theorie der Opaleszenz von homogenen Flüssigkeiten und Flüssigkeitsgemischen in der Nähe des kritischen Zustandes. *Annalen der Physik* 33: 1275–1298.

Elbert, W., et al. 2012. Contributions of cryptogamic covers to the global cycles of carbon and nitrogen. *Nature Geoscience* 5: 459–462.

Erickson, G. 1981. Geology and origin of the Chilean nitrate deposits. *United States Geological Survey Professional Paper* 1188.

Erisman, J. W., et al. 2008. How a century of ammonia synthesis changed the world. *Nature Geoscience* 1: 636–639.

Fields, S. 2004. Global nitrogen: Cycling out of control. *Environmental Health Perspectives* 112: A556–A563.

Finney B. P., I. Gregory-Eaves, M. S. V. Douglas, and J. P. Smol. 2002. Fisheries productivity in the northeastern Pacific Ocean over the past 2,200 years. *Nature* 416: 729–732.

———, et al. 2000. Impacts of climatic change and fishing on Pacific salmon abundance over the past 300 years. *Science* 290: 795–797.

Fogel, M. L., N. Tuross, B. J. Johnson, and G. H. Miller. 1997. Biogeochemical record of ancient humans. *Organic Geochemistry* 27: 275–287.

Fryzuk, M. D. 2004. Inorganic chemistry: Ammonia transformed. *Nature* 427: 498–499.

Fuller, B. T., et al. 2004. Nitrogen balance and $\delta^{15}N$: Why you're not what you eat during pregnancy. *Rapid Communications in Mass Spectrometry* 18: 2889–2896.

Goran, M. 1947. The present-day significance of Fritz Haber.

American Scientist 35: 400–403. http://soils.wisc.edu/facstaff/
barak/soilscience326/haber_amsci.htm.

Gruber, N., and J. N. Galloway. 2008. An Earth-system perspective
of the global nitrogen cycle. *Nature* 451: 293–296.

Hilderbrand, G. V., et al. 1996. Use of stable isotopes to determine diets
of living and extinct bears. *Canadian Journal of Zoology* 74: 2080–2088.

Hines, S. 2011. Nitrogen from humans pollutes remote lakes for more
than a century. *University of Washington, News and Information*, 19
Dec. 2011. http://www.washington.edu/news/2011/12/15/nitrogen
-from-humans-pollutes-remote-lakes-for-more-than-a-century/.

Hocking, M. D., and J. D. Reynolds. 2011. Impacts of salmon on
riparian plant diversity. *Science* 331: 1609–1612.

Holtgrieve, G. W., et al. 2011. A coherent signature of anthropogenic
nitrogen deposition to remote watersheds of the Northern Hemisphere. *Science* 334: 1545–1548.

Howarth, R. W., E. W. Boyer, W. J. Pabich, and J. N. Galloway. 2002.
Nitrogen use in the United States from 1961–2000 and potential
future trends. *Ambio* 31: 88–96.

King, G. 2012. Past imperfect: Fritz Haber's experiments in life and
death. Smithsonian blog (posted June 6, 2012). http://blogs.
smithsonianmag.com/history/2012/06/fritz-habers-experiments
-in-life-and-death/.

Law, B. 2013. Nitrogen deposition and forest carbon. *Nature* 496:
307–308.

Liu, X., et al. 2013. Enhanced nitrogen deposition over China.
Nature 494: 459–462.

Locke, S. 2001. Ancient salmon hold clues to cycles. *yourYukon*,
Column 245, Series 1. http://www.taiga.net/yourYukon.col245
.html.

Nelson, D. E., et al. 2012. An isotopic analysis of the diet of the
Greenland Norse. *Journal of the Atlantic* 3: 93–118.

O'Connell, T. C., C. J. Kneale, and G. C. C. Kuhnle. 2012. The
diet-body offset in human nitrogen isotopic values: A controlled
dietary study. *American Journal of Physical Anthropology* 149:
426–434.

Pittmann, R. 2009. An odd story from America's Civil War: Gunpowder. Bard of the South blog, March 1, 2009. http://www.bardofthesouth.com/an-odd-story-from-americas-civil-war-gunpowder/.

Poulin, L. 2006. Estimating volumes of air through various engines in an urban setting. *CMOS Bulletin SCMO* 34: 116–124.

Reimchen, T. 2001. Salmon nutrients, nitrogen isotopes, and coastal forests. *Ecoforestry* 16: 13–17.

Sheppard, D. 2009. Robert Le Rossignol, 1884–1976, professional chemist. *ChemUCL Newsletter*. http://www.ucl.ac.uk/chemistry/alumni/documents/A5booklet_020909.pdf.

Sobel'man, I. I. 2002. On the theory of light scattering in gases. *Physics—Uspekhi* 45: 75–80.

Smoluchowski, M. 1908. Zur kinetischen Theorie der Brownschen Molekularbewegung und der Suspensionen. *Annalen der Physik* 25: 205.

Stevenson, A. Speech, "The Atomic Future" (18 Sept. 1952). Quoted in *A. Stevenson, Speeches* (1953). Edited by Richard Harrity. 129.

Townsend, A. R., et al. 2003. Human health effects of a changing global nitrogen cycle." *Frontiers in Ecology and the Environment* 1: 240–246.

Vitousek, P. M., et al. 1997. Human alteration of the global nitrogen cycle: Causes and consequences. *Ecological Applications* 7: 737–750.

Zwanzig, R. 1964. On the validity of the Einstein-Smoluchowski Theory of light scattering. *Journal of the American Chemical Society* 86: 3489–3493.

CHAPTER 7: BONES AND STONES

Adamo, P., and P. Violante. 2000. Weathering of rocks and neogenesis of minerals associated with lichen activity. *Applied Clay Science* 16: 229–256.

Allen, P. A. 2008. From landscapes into geological history. *Nature* 451: 274–276.

Babikova, Z.,et al. 2013. Underground signals carried through common mycelial networks warn neighboring plants of aphid attack. *Ecology Letters* 16: 835–843.

Bais, H. P., et al. 2006. The role of root exudates in rhizosphere interactions with plants and other organisms. *Annual Review of Plant Biology* 57: 233–266.

Banfield, J. F., W. W. Barker, S. A. Welch, and A. Taunton. 1999. Biological impact on mineral dissolution: Application of the lichen model to understanding mineral weathering in the rhizosphere. *Proceedings of the National Academy of Sciences* 96: 3404–3411.

Behie, S. W., P. M. Zelisko, and M. J. Bidochka. 2012. Endophytic insect-parasitic fungi translocate nitrogen directly from insects to plants. *Science* 336: 1576–1577.

Blum, J. D., et al. 2002. Mycorrhizal weathering of apatite as an important calcium source in base-poor forest ecosystems. *Nature* 417: 729–731.

Boskey, A. L. 2007. Mineralization of bones and teeth. *Elements* 3: 387–393.

Brown, F., J. Harris, R. Leakey, and A. Walker. 1985. Early *Homo erectus* skeleton from west Lake Turkana, Kenya. *Nature* 316: 788–792.

Carpenter, K. 2005. Experimental investigation of the role of bacteria in bone fossilization. *Neues Jahrbuch für Geologie und Paläontologie Monatshefte* 2: 83–94.

Carpenter, K. How to make a fossil: Part 1—Fossilizing bone. *Journal of Paleontological Sciences*, JPS.C.07.0001.

Chen, J., H.-P. Blume, and L. Beyer. 2000. Weathering of rocks induced by lichen colonization—a review. *Catena* 39: 121–146.

Daniel, J. C., and K. Chin. 2010. The role of bacterially mediated precipitation in the permineralization of bone. *Palaios* 25: 507–516.

Dyke, J. G., F. Gans, and A. Kleidon. 2011. Towards understanding how surface life can affect interior geological processes: A non-equilibrium thermodynamics approach. *Earth System Dynamics* 2: 139–160.

Ferguson, B. A., et al. 2003. Coarse-scale population structure of pathogenic *Armillaria* species in a mixed-conifer forest in the Blue Mountains of northeast Oregon. *Canadian Journal of Forest Research* 33: 612–623.

Frezzotti, M. L., J. Selverstone, Z. D. Sharp, and R. Compagnoni. 2011. Carbonate dissolution during subduction revealed by diamond-bearing rocks from the Alps. *Nature Geoscience* 4: 703–706.

Guidry, M. W., and F. T. Mackenzie. 2003. Experimental study of igneous and sedimentary apatite dissolution: Control of pH, distance from equilibrium, and temperature on dissolution rates. *Geochimica et Cosmochimica Acta* 67: 2949–2963.

Hazen, R. M., et al. 2008. Mineral evolution. *American Mineralogist* 93: 1693–1720.

———. 2010. Evolution of minerals. *Scientific American*, March 2010, pp. 58–65.

Heckman, D. S., et al. 2001. Molecular evidence for the early colonization of land by fungi and plants. *Science* 293: 1129–1133.

Hinsinger, P., A. G. Bengough, D. Vetterlein, and I. M. Young. 2009. Rhizosphere: Biophysics, biogeochemistry and ecological relevance. *Plant Soil* 321: 117–152.

Jones, D. L., C. Nguyen, and R. D. Finlay. 2009. Carbon flow in the rhizosphere: Carbon trading at the soil-root interface. *Plant Soil* 321: 5–33.

Kiers, E. T., et al. 2011. Reciprocal rewards stabilize cooperation in the mycorrhizal symbiosis. *Science* 333: 880–882.

———, R. A. Rousseau, S. A. West, and R. F. Denison. 2003. Host sanctions and the legume-rhizobium mutualism. *Nature* 425: 78–81.

Kump, L. R. 2008. The rise of atmospheric oxygen. *Nature* 451: 277–278.

Lambers, H., C. Mougel, B. Jaillard, and P. Hinsinger. 2009. Plant-microbe-soil interactions in the rhizosphere: An evolutionary perspective. *Plant Soil* 321: 83–115.

Landeweert, R., et al. 2001. Linking plants to rock, ectomycorrhizal fungi mobilize nutrients from minerals. *Trends in Ecology and Evolution* 16: 248–253.

Lenton, T. M. 2001. The role of land plants, phosphorus weathering and fire in the rise and regulation of atmospheric oxygen. *Global Change Biology* 7: 613–629.

Leopold, A. 1949. *A Sand County Almanac: And Sketches Here and There*. Oxford University Press, NY.

Malina, R. M. 2005. In *Human Body Composition*. 2nd ed. Edited by S. Heymsfield, T. Lohman, Z. Wang, and S. Going, 271–298. Human Kinetics, Champaign.

Manolagas, S. C. 2000. Birth and death of bone cells: Basic regulatory mechanisms and implications for the pathogenesis and treatment of osteoporosis. *Endocrine Reviews* 21: 115–137.

Milgrom, C., et al. 2000. Using bone's adaptation ability to lower the incidence of stress fractures. *American Journal of Sports Medicine* 28: 245–251.

Mitchell, H. H., T. S. Hamilton, F. R. Steggerda, and H. W. Bean. 1945. The chemical composition of the adult human body and its bearing on the biochemistry of growth. *Journal of Biological Chemistry* 158: 625–637.

Rodell, M., et al. 2005. Global biomass variation and its geodynamic effects: 1982–98. *Earth Interactions* 9: 1–19.

Rosing, M. T. 2008. On the evolution of minerals. *Nature* 456: 456–458.

Selosse, M.-A., and F. Roussel. 2011. The plant-fungal marketplace. *Science* 333: 828–829.

Smith, S. E., I. Jakobsen, M. Grønlund, and F. A. Smith. 2011. Roles of arbuscular mycorrhizas in plant phosphorus nutrition: Interactions between pathways of phosphorus uptake in arbuscular mycorrhizal roots have important implications for understanding and manipulating plant phosphorus acquisition. *Plant Physiology* 156: 1050–1057.

Sobel, A. E., M. Rockenmacher, and B. Kramer. 1945. Carbonate content of bone in relation to the composition of blood and diet. *Journal of Biological Chemistry* 158: 475–489.

Song, Y. Y., et al. 2010. Interplant communication of tomato plants through underground common mycorrhizal networks. PLOS ONE 5: e13324. doi:10.1371/journal.pone.0013324.

Techawiboonwong, A., H. K. Song, M. B. Leonard, and F. W. Wehrli. 2008. Cortical bone water: *In vivo* quantification with ultrashort echo-time MR imaging. *Radiology* 248: 824–833.

Timmins, P. A., and J. C. Wall. 1977. Bone water. *Calcified Tissue Research* 23: 1–5.

Wallach, S., and C. D. Berdanier. 2008. Macromineral nutrition, disorders of skeletons and kidney stones. In *Handbook of Nutrition and Food, Second Edition*. Edited by C. D. Berdanier, J. Dwyer, and E. B. Feldman, 1079–1092. CRC Press, Boca Raton.

Walter, M. J., et al. 2011. Deep mantle cycling of oceanic crust: Evidence from diamonds and their mineral inclusions. *Science* 334: 54–57.

Wopenka, B., and J. D. Pasteris. 2005. A mineralogical perspective on the apatite in bone. *Materials Science and Engineering* 25: 131–143.

CHAPTER 8: LIMITS TO GROWTH

Adams, F. C., and G. Laughlin. 1997. A dying universe: The long-term fate and evolution of astrophysical objects. *Reviews of Modern Physics* 69: 337–372.

Annunziato, A. T. 2008. DNA packaging: Nucleosomes and chromatin. *Nature Education* 1: 26.

Asimov, I. 1962. *Fact and Fancy*. Doubleday and Company.

Ben-Ami, Y., et al. 2010. Transport of North African dust from the Bodélé depression to the Amazon basin: A case study. *Atmospheric Chemistry and Physics Discussions* 10: 4345–4372.

Berkow, R. 1997, ed. *Merck Manual of Medical Information*. Merck Research Laboratories.

Bianconi, E., et al. 2013. An estimation of the number of cells in the human body. *Annals of Human Biology* 6: 463–471.

Bird, J., et al. 1993. Evidence for correlated changes in the spectrum and composition of cosmic rays at extremely high energies. *Physical Review Letters* 71: 3401–3404.

Bouwman, L., et al. 2011. Exploring global changes in nitrogen and phosphorus cycles in agriculture induced by livestock production over the 1900–2050 period. *Proceedings of the National Academy of Sciences, Early Edition*. doi: 10.1073/pnas.1012878108.

Bradbury, J. P., S. M. Colman, and R. L. Reynolds. 2004. The history of recent limnological changes and human impact on Upper Klamath Lake, Oregon. *Journal of Paleolimnology* 31: 151–165.

Bristow, C. S., K. A. Hudson-Edwards, and A. Chappell. 2010. Fertilizing the Amazon and equatorial Atlantic with West African dust. *Geophysical Research Letters* 37, L14807, doi:10.1029/2010GL043486.

Casewatch. 2007. Cell Tech sued for wrongful death. (Posted March 16, 2007.) http://www.casewatch.org/civil/celltech/complaint.shtml.

Childers, D. L., J. Corman, M. Edwards, and J. J. Elser. 2011. Sustainability challenges of phosphorus and food: Solutions from closing the human phosphorus cycle. *BioScience* 61: 117–124.

Clabby, S. 2010. Does peak phosphorus loom? *American Scientist* 98: 291.

Conley, D. J. 2012. Save the Baltic Sea. *Nature* 486: 463–464.

Cordell, D., and S. White. 2011. Peak phosphorus: Clarifying the key issues of a vigorous debate about long-term phosphorus security. *Sustainability* 3: 2027–2049.

Darwin, C. 1860. *A Naturalist's Voyage Around the World*. John Murray.

Deutsch, C., and T. Weber. 2012. Nutrient ratios as a tracer and driver of ocean biogeochemistry. *Annual Review of Marine Science* 4: 113–141.

Elser, J., and S. White. 2010. Peak phosphorus. *Foreign Policy* (accessed April 11, 2014). http://www.foreignpolicy.com/articles/2010/04/20/peak_phosphorus#sthash.egcYpLhd.dpbs.

Elser, J., and E. Bennett. 2011. A broken biogeochemical cycle. *Nature* 478: 29–31.

Evans, R. 2002. Blast from the past. *Smithsonian Magazine,* July 2002. http://www.smithsonianmag.com/history-archaeology/blast.html.

Formenti, P., et al. 2008. Regional variability of the composition of mineral dust from western Africa: Results from the AMMA SOP0/DABEX and DODO field campaigns. *Journal of Geophysical Research* 113, D00C13, doi: 10.1029/2008JD009903.

Freitas, R. A. 1998. Nanomedicine. http://www.foresight.org/Nanomedicine/Ch03_1.html.

Gilbert, N. 2009. The disappearing nutrient. *Nature* 461: 716–718.

Giles, J. 2005. The dustiest place on Earth. *Nature* 434: 816–819.

Gilroy, D. J., et al. 2000. Assessing potential health risks from microcystin toxins in blue-green algae dietary supplements. *Environmental Health Perspectives* 108: 435–439.

Gross, A., D. Castido, C. Pio, and A. Angert. 2013. African dust phosphorus fertilizing Amazon and the Atlantic Ocean is derived from marine sediments and igneous rocks—no indication for Bodélé diatomite contribution. *Geophysical Research Abstracts* 15, EGU2013–2060.

Heussner, A. H., L. Mazija, J. Fastner, and D. R. Dietrich. 2012. Toxin content and cytotoxicity of algal dietary supplements. *Toxicology and Applied Pharmcology* 265: 263–271.

Hietz., P., et al. 2011. Long-term change in the nitrogen cycle of tropical forests. *Science* 334: 664–666.

International Geological Correlation Programme. 1989. In *Phosphate Deposits of the World.* Vol. 2, *Phosphate Rock Resources.* Edited by A. J. G. Nothold, R. P. Sheldo, and D. F. Davidson. Cambridge University Press, UK.

Jasinski, S. 2013. Phosphate rock. *United States Geological Survey Mineral Commodity Summaries.* http://minerals.er.usgs.gov/minerals/pubs/commodity/phosphate_rock/mcs-2013-phosp.pdf.

Kerr, R. A. 2011. Peak oil production may already be here. *Science* 331: 1510–1511.

———. 2012. Are world oil's prospects not declining all that fast? *Science* 337: 633.

Kolbert, E. 2011. Billions and billions. *The New Yorker* (Oct. 24, 2011), pp. 21–22.

Koo, B.-C., et al. 2013. Phosphorus in the young supernova remnant Cassiopeia A. *Science* 342: 1346–1348.

Koren, I., Y. J. Kaufman, R. Washington, M. C. Todd, Y. Rudich, J. V. Martins, and D. Rosenfeld. 2006. The Bodélé depression: A single spot in the Sahara that provides most of the mineral dust to the Amazon forest. *Environmental Research Letters*. doi:10.1088/1748-9326/1/1/014005.

Lancaster, K. M., et al. 2011. X-ray emissions spectroscopy evidences a central carbon in the nitrogenase iron-molybdenum cofactor. *Science* 334: 974–977.

Liebig, J. 1851. *Familiar Letters on Chemistry in Its Relations to Physiology, Dietetics, Agriculture, Commerce and Political Economy*. Taylor, Walton, and Maberly.

Lougheed, T. 2011. Phosphorus paradox: Scarcity and overabundance of a key nutrient. *Environmental Health Perspectives* 119: A209–A213.

Mahmood, N. A., and W. W. Carmichael. 1986. Paralytic shellfish poisons produced by the freshwater cyanobacterium *Aphanizomenon flos-aquae* NH-5. *Toxicon* 24: 175–186.

Malina, R. M. 2005. In *Human Body Composition*. 2nd ed. Edited by S. Heymsfield, T. Lohman, Z. Wang, and S. Going, 271–298. Human Kinetics, Champaign. Mayo Clinic Staff, 2011. Complete blood count (CBC). (Posted Jan. 29, 2011.) http://www.mayoclinic.com/health/complete-blood-count/MY00476/DSECTION=results.

Murray, R. W., M. Leinen, and C. W. Knowlton. 2012. Links between iron input and opal deposition in the Pleistocene equatorial Pacific Ocean. *Nature Geoscience* 5: 270–274.

Owen, J. 2010. World's largest dead zone suffocating sea. *National Geographic News* (March 5, 2010). http://news.nationalgeographic.com/news/2010/02/100305-baltic-sea-algae-dead-zones-water/.

Pennisi, E. 2012. Water reclamation going green. *Science* 337: 674–676.

Redeker, I. H. 1976. Beneficiation and conservation of North Carolina phosphates. Transcript of presentation at Centennial Meeting of the American Chemical Society in New York, April 7, 1976.

Revkin, A. C. 2009. Fertilizer divide: Too much, not enough. Dot Earth blog, *New York Times* (June 19, 2009). http://dotearth.blogs.nytimes.com/2009/06/19/fertilizer-divide-too-much-not-enough/?_r=0.

Richier, S., et al. 2012. Abundances of iron-binding photosynthetic and nitrogen-fixing proteins of *trichodesmium* both in culture and *in situ* from the North Atlantic. PLOS ONE 7:e35571.

Rolfe, D. F., and G. C. Brown. 1997. Cellular energy utilization and molecular origin of standard metabolic rate in mammals. *Physiological Reviews* 77: 731–758.

Rubin, M., I. Berman-Frank, and Y. Shaked. 2011. Dust- and mineral-iron utilization by the marine dinitrogen-fixer *Trichodesmium*. *Nature Geoscience* 4: 529–534.

Schindler, D. W. 1974. Eutrophication and recovery in experimental lakes: Implications for lake management. *Science* 184: 897–899.

Schindler, D. W., et al. 2008. Eutrophication of lakes cannot be controlled by reducing nitrogen input: Results of a 37-year whole-ecosystem experiment. *Proceedings of the National Academy of Sciences* 105: 11,254–11,258.

Schröder, J. J., D. Cordell, A. L. Smith, and A. Rosemarin. 2010. Sustainable use of phosphorus. *Report 357*, Plant Research International, Wageningen University and Research Center, the Netherlands.

Smil, V. 2000. Phosphorus in the environment: Natural flows and human interferences. *Annual Review of Energy and the Environment* 25: 53–88.

Törnroth-Horsefield, S., and R. Neutze. 2008. Opening and closing the metabolite gate. *Proceedings of the National Academy of Sciences* 105: 19,565–19,566.

Townsend, A. R., et al. 2003. Human health effects of a changing global nitrogen cycle. *Frontiers in Ecology and the Environment* 1: 240–246.

Trenberth, K. E., and L. Smith. 2005. The mass of the atmosphere: A constraint on global analyses. *Journal of Climate* 18: 864–875.

Van Kauwenbergh, S. J. 2010. *World Phosphate Rock Reserves and Resources.* International Fertilizer Development Center (IFDC), Muscle Shoals.

Vogel, G. 2012. Finding a new way to go. *Science* 337: 673.

Washington, R., et al. 2009. Dust as a tipping element: The Bodélé depression, Chad. *Proceedings of the National Academy of Sciences* 106: 20,564–20,571.

Whitsett, D. 2013. Environmentalist claims about Upper Klamath Lake fall short. (From Senator Whitsett's August 7 newsletter.) *Klamath News.net.* http://klamathnews.net/senator-doug-whitsett/2013/8/10/environmentalist-claims-about-upper-klamath-lake-fall-short.

Chapter 9: Fleeting Flesh

Adams, F. C., and G. Laughlin. 1997. A dying universe: The long-term fate and evolution of astrophysical objects. *Reviews of Modern Physics* 69: 337–372.

Aebersold, P. C. 1954. Radioisotopes—New keys to knowledge. *Annual Report of the Board of Regents of the Smithsonian Institution 1953*, 219–240. Publication 4149, showing the operations, expenditures and condition of the Institution for the year ended June 30, 1953. U.S. Government Printing Office. http://www.archive.org/stream/annualreportofbo1953smit/annualreportofbo1953smit_djvu.txt.

Asimov, I. 1955. The radioactivity of the human body. *Journal of Chemical Education* 32: 84–85.

Benner, S. A. Defining life. *Astrobiology* 10: 1021–1030.

Bergmann, O., et al. 2009. Evidence for cardiomyocyte renewal in humans. *Science* 324: 98–102.

———, and J. Frisén. 2013. Why adults need new brain cells. *Science* 340: 695–696.

Bianconi, E., et al. 2013. An estimation of the number of cells in the human body. *Annals of Human Biology* 40: 463–471.

Brock, W. H. 1997. *Justus von Liebig: The Chemical Gatekeeper.* Cambridge University Press, UK.

Bryson, B. 2003. *A Short History of Nearly Everything.* Broadway Books.

Burley, M. W., E. M. Ritchie, and C. R. Gray. 1957. Winds of the upper troposphere and lower stratosphere over the United States. *Monthly Weather Review* 85: 11–17.

Canham, P. B., and A. C. Burton. 1968. Distribution of size and shape in populations of normal human red cells. *Circulation Research* 22: 405–422.

Chen, M., et al. 2005. Hepatocytes express abundant surface Class I MHC and efficiently use transporter associated with antigen processing, tapasin, and low molecular weight polypeptide proteasome subunit components of antigen processing and presentation pathway. *Journal of Immunology* 175: 1047–1055.

Clarke, B. 2008. Normal bone anatomy and physiology. *Clinical Journal of the American Society of Nephrology* 3: 131–139.

Clery, D. 2012. What's the source of the most energetic cosmic rays? *Science* 336: 1096–1097.

Conover, W. C., and C. J. Wentzien. 1955. Winds and temperatures to forty kilometers. *Journal of Meteorology* 12: 160–164.

Creamer, B., R. G. Shorter, and J. Bamforth. 1961. The turnover and shedding of epithelial cells. *Gut* 2: 110–116.

Dawkins, R. 2006. *The God Delusion.* Mariner Books.

Deftos, L. J. 2010. Calcium and phosphate homeostasis. ENDOTEXT. http://www.endotext.org/parathyroid/parathyroid2/indx.html.

Doherty, M. K., et al. 2009. Turnover of the human proteome: Determination of protein intracellular stability by dynamic SILAC. *Journal of Proteome Research* 8: 104–112.

Dorneanu, L. 2007. Why ghosts can't go through walls . . . and more-scientific arguments. *Softpedia* (posted May 26, 2007). http://news.softpedia.com/news/Why-Ghosts-Can-039-t-Go-Through-Walls-and-More-Scientific-Arguments-55697.shtml/.

Djuricin, S., X. Xu, and D. E. Pataki. 2012. The radiocarbon composition of tree rings as a tracer of fossil fuel emissions in the Los Angeles basin: 1980–2008. *Journal of Geophysical Research Atmospheres* 117, D12,27. doi: 10.1029/2011JD017284.

Forster, L., et al. Natural radioactivity and human mitochondrial DNA mutations. *Proceedings of the National Academy of Sciences* 99: 13,950–13,954.

Freitas, R. A. 1998. Nanomedicine. http://www.foresight.org/ Nanomedicine/Ch03_1.html. *Guidebook* 2006. EMEP/Corinair Emission Inventory Guidebook, version 4.

European Environmental Agency. Technical report No 11/2006.

Hawking, S. 2007. *A Stubbornly Persistent Illusion: The Essential Scientific Works of Albert Einstein.* Running Press.

Heinemeier, K. M., et al. 2013. Lack of tissue renewal in human adult Achilles tendon is revealed by nuclear bomb ^{14}C. *FASEB Journal* 27: 2074–2079.

Joseph, J. A., S. Erat, and B. M. Rabin. 1998. CNS effects of heavy particle irradiation in space: Behavioral implications. *Advances in Space Research* 22: 209–216.

Kestenbaum, D. 2007. Atomic tune-up: How the body rejuvenates itself. National Public Radio. (Published July 14, 2007.) http://m .npr.org/story/11893583.

Leiper, J. B., A. Carnie, and R. J. Maughan. 1996. Water turnover rates in sedentary and exercising middle aged men. *British Journal of Sports Medicine* 30: 24–26.

Lindemann, B. 2001. Receptors and transduction in taste. *Nature* 413: 219–225.

Lynnerup, N., et al. 2008. Radiocarbon dating of the human eye lens crystallines reveal proteins without carbon turnover throughout life. *PLOS ONE* 3:e 1529. doi: 10.1371/journal. pone.0001529.

Masarik, J., and J. Beer. 2009. An updated simulation of particle fluxes and cosmogenic nuclide production in the Earth's atmosphere. *Journal of Geophysical Research* 114. doi: 10.1029/ 2008JD010557.

Malina, R. M. 2005. In *Human Body Composition.* 2nd ed. Edited by

S. Heymsfield, T. Lohman, Z. Wang, and S. Going, 271–298. Human Kinetics, Champaign.

Misra, M. K., K. W. Ragland, and A. J. Baker. 1993. Wood ash composition as a function of furnace temperature. *Biomass and Bioenergy* 4: 103–116.

Murry, C. E., and R. T. Lee. 2008. Turnover after the fallout. *Science* 324: 47–48.

Pawelczyk, S., and A. Pazdur. 2004. Carbon isotopic composition of tree rings as a tool for biomonitoring CO_2 level. *Radiocarbon* 46: 701–719.

Peterson, B. J., and B. Fry. 1987. Stable isotopes in ecosystem studies. *Annual Reviews of Ecology and Systematics* 18: 293–320.

Pollycove, M., and R. Mortimer. 1961. The quantitative determination of iron kinetics and hemoglobin synthesis in human subjects. *Journal of Clinical Investigation* 40: 753–782.

Ramesh, N., M. Hawron, C. Martin, and A. Bachri. 2011. Flux variation of cosmic muons. *Journal of the Arkansas Academy of Science* 65: 1–5.

Rozenzweig, A. 2012. Cardiac regeneration. *Science* 338: 1549–1550.

Schrier, S. L. 2013. Red blood cell survival: Normal values and measurement. *UpToDate.* http://www.uptodate.com/contents/red-blood-cell-survival-normal-values-and-measurement. (Accessed Dec. 22, 2013.)

Schutz, Y. 2011. Protein turnover, ureagenesis and gluconeogenesis. *International Journal for Vitamin and Nutritional Research* 81: 101–107.

Seale, R. U. 1959. The weight of the dry fat-free skeleton of American whites and negroes. *American Journal of Physical Anthropology* 17: 37–48.

Shepard, I. 1956. *Jack London's Tales of Adventure.* Doubleday.

Spalding, K. L., et al. 2005. Retroactive dating of cells. *Cell* 122: 133–143.

———. 2005. Age written in teeth by nuclear tests. *Nature* 437: 333–334.

———, et al. 2006. Dynamics of fat cell turnover in humans. *Nature* 453: 783–787.

Terjung, R. L. 1979. The turnover of cytochrome c in different skeletal-muscle fibre types of the rat. *Biochemistry Journal* 178: 569–574.

Tirziu, D., F. J. Giordano, and M. Simons. 2010. Cell communications in the heart. *Circulation* 122: 928–937.

Trotter, M., and B. B. Hixon. 1974. Sequential changes in weight, density, and percentage ash weight of human skeletons from an early fetal period through old age. *Anatomical Record* 179: 1–18.

United States Environmental Protection Agency. 2013. Calculate your radiation dose. http://www.epa.gov/radiation/understand/calculate.html.

University of Utah. 2004. Cosmic ray observatory planned. *U News Center.* (January 12, 2004.) http://unews.utah.edu/old/p/030906-24.html.

Vass, A. A. 2001. Beyond the grave—understanding human decomposition. *Microbiology Today* 28: 190–192.

Wade, N. 2005. Your body is younger than you think. *New York Times* (August 2). http://www.nytimes.com/2005/08/02/science/02cell.html?ei=5058&en=d9f7fc175dea3641&ex=1123646400&partner=IWON&pagewanted=print&_r=0.

Wang, Z., et al. 2003. Total body protein: A new cellular level mass and distribution prediction model 1'2'3. *American Journal for Clinical Nutrition* 78: 979–984.

Watson, A. A. 2002. Extensive air showers and ultra high energy cosmic rays. Lecture given at a summer school in Mexico. http://www.ast.leeds.ac.uk/Auger/augerthesis/mexlects3.pdf.

Weinstein, G. D., J. L. McCullough, and P. Ross. 1984. Cell proliferation in normal epidermis. *Journal of Investigative Dermatology* 82: 623–628.

Welle, S. 1999. *Human Protein Metabolism.* Springer.

Yaemsiri, S., N. Hou, M. M. Slining, and K. He. 2010. Growth rate of human fingernails and toenails in healthy American young adults. *Journal of the European Academy of Dermatology and Venereology* 24: 420–423.

Epilogue: Einstein's Adirondacks

Brown, P. 2007. Wilderness advocate. *New York State Conservationist* 61: 2–6.

Capra, F. 2002. *The Hidden Connections: A Science for Sustainable Living*. Random House.

Einstein, A. 1932. My credo. *Albert Einstein Archiv*. Call Nr 28-218.00. Original and translated text of handwritten letter posted online at www.einstein-Web site.de/z_biography/credo.html.

Foster, B. 2005. Einstein and his love of music. *Physics World* (January 2005): 1.

Isaacson, W. 2008. *Einstein: His Life and Universe*. Simon & Schuster.

Miller, A. I. 2006. A genius finds inspiration in the music of another. *New York Times*, January 31.

Schipler, D. K. 1983. *Russia: Broken Idols, Solemn Dreams*. Crown Publishing Group.

White, P. 2005. Albert Einstein: The violinist. *The Physics Teacher* 43: 286–288.

Whitman, W. 1881–1882. *Leaves of Grass*. James R. Osgood and Company.

Zartman, B. 2013. Einstein's boats. *BoatUS Magazine* (April/May 2013): 62–63.

Index

Abe Lincoln's axe, 213
acid rain, 238
adenosine triphosphate (ATP), 195–96
Adirondack Mountains, Einstein's
    summer retreat in, 233–44
Aebersold, Paul, 213–14
Africa, human origins in, 158–65
air
    composition of, 15, 19, 22–26, 32, 87,
        97–100, 133
    oscillations in composition of,
        23–24, 32
    See also atmosphere
air pollution, 17, 32, 86, 238
air pressure, at sea level and at
        altitudes, 27–28
air travel
    air supply during, 28
    radiation during, 217
Alaskan Native Americans, 75
alchemists, and science, 15
alfalfa, 83, 138
algae, 27, 83
    as dietary supplement, 182–83
algal blooms, 185
Allen, Philip, 178
altitude
    and air pressure, 27–28
    and radiation, 217

alveoli, 19
Amazon rain forests, 190
American Civil War, 143
ammonia
    biological, 138, 143, 155
    industrial production, 141–47,
        155
ammonium nitrate, 141
Andeans, 28, 115
anemia, 76
Anthropocene epoch (age of humans),
        6–7
antibiotics, 127
anti-Semitism, 141
apatite
    biological, in skeleton, 162–63, 165,
        167, 178
    mineral, 174–75, 201
Aphanizomenon, 186–87
argon, 33–35
ashes of cremation, 223
    dispersal of atoms of, 226–27
"ashes to ashes, dust to dust," 220
Asimov, Isaac, 218
Aston, Francis, 240
astronauts, 46, 168
athletes, 127
    skeletons of, 167–68
Atlanta, Georgia, 26

atmosphere
  humidity of, 43–44
  radiation of light from, 135
  scattering of light in, 136–37
  total mass of, 33
  water evaporated into, 57
  See also air
atom bomb, 239–40
atomic power, 239
atomic theory
  ancient Greek, 3, 40
  modern, 40
atoms
  number of, in the human body,
    197–98
  recycling of, in the human body,
    91–93, 206–15
  seen in an electron microscope, 8
  structure of, 20
  thermal motion of, 41–42
  your body composed of, 1–2, 5, 9
Aurora, North Carolina, phosphate
    deposits, 200–202
Australia, 107
Austria, 106

bacteria
  body's defenses against, 82, 127
  fossilization of, 101
  manganese in some types of, 83
  nitrogen-fixing by, 138
Balsam, Artur, 243
Baltic Sea, 187, 194
Bänziger, Hans, 110–12, 116
bats, 114
bears, 149–51
Becher, Johann, 15
bees, 135
Behlen, Charles, 74
belemnite, 104–5
bends, the, 134
Bergmann, Olaf, 210

Bianconi, Eva, 210
Big Bang, 30, 37, 230
bilirubin, 78
biofuels, 196
biominerals, 176–79
birth defects, 218
bitter taste, 113, 122
Blake, Melissa, 182, 187
blood
  color of, 75–76, 77
  composition of, 208–9
  iron in, 76–82
  journey of, through the body, 52–53
  oxygen in, 19, 78–80
  See also red blood cells
blue light, 135
blue people, 74–76
Blum, Joel, 173
Bodélé Depression, 189–92
Bohren, Craig, 136
bone marrow, regeneration of cells in,
    212
bones, 163–68
  alive nature of, 163
  broken, self-repair of, 167
  dispersal of atoms from, after death,
    222–23
  function of, 162
  regeneration of cells in, 212
  structure of, 14, 163–67
Bosch, Carl, 142
Bowen, Gabriel, 48
Boyer, George, 17
Boyle, Robert, 15
brain, regeneration of cells in, 211
Brazil, 106
breath
  final, 13
  of historical figures, 32–35
  water in, 44
breathing, process of, 11–14, 19–22,
    93–94, 133–34
bronchioles, 19

Brown, Frank, 159–60
Brown, Robert, 37
Brownian motion, 37–42
  cost-free within cells, 51–53
Bryson, Bill, 213
bubonic plaque, 82
burial ceremonials, Anglican, 220
burning, of carbon compounds, 25,
  91–92, 97–98
butterflies, 110–11

calcium
  dispersal from body after death, 227
  formation of, in stars, 71
  in the nervous system, 129
  in plants, 169, 173, 174–75
  in the skeleton, 162–63, 167, 171
California, 104
capillaries, 19
capsaicin, 130
carbon
  amount of, in the human body,
    88–89
  atomic structure of, 89–90
  chemical bonds formed by, 89–90
  dispersal from body after death,
    221–22, 224, 225–26
  emissions from human activity, 6–7
  formation of, in stars, 71
  in human body, 6, 88–89, 96–97, 198
  from plants, 169
  in steel, 63
carbon-13, 105, 107, 226
carbon-14, 217–19, 226, 240
  as isotopic tracer, 210
carbon compounds, 90–93
  unmaking of, 91–93
carbon dioxide
  in the air, 16, 23–26, 32, 94–100, 199,
    239
  in breathing process, 19–22, 93–94
  burial of, 178

in oceans, 89
rising concentration of, in
  atmosphere, 24–25, 87, 97–100
Carbon Dioxide Information Analysis
  Center, 99
CarbonTracker Web site, 98
Caribbean islands, 190
cartilage, 166–67
catalase, 80–81
Cawein, Madison, 75–76
cells
  Brownian motion in, 51–53
  death of, 221
  machinery of, 20–21
  membrane surrounding, 193–95
  number of, in the human body, 210
  salt in, 125–26
  variety of shape and size of, 210
Cell Tech, 182
Cerling, Thure, 54
cesium-137, 240
Chaptal, Jean-Antoine, 143
chemical bonds, 4–5
children, immortality hoped from, 230
Chile, 142
China, ancient, science in, 60–61,
  64–65
China, modern
  industry in, 84
  phosphate resources, 204
chlorine
  dispersal from body after death, 222
  formation of, in stars, 71
  in the nervous system, 129
  as poison gas, 121, 144–45
  radioactive, in rocks, 219
  in salt, 113, 119
  source of, in nature, 118, 120
chlorophylls, 29, 140, 150
chloroplasts, 29
Christensen, Jennie, 150
chrysanthemums, 130
Churchill, Winston, 146

cities
  ecosystems of, 94–97
  nitrous oxide generated from,
    140–41
clay, eating of (geophagy), 114–15
coal
  geologic origin of, 100–101, 178
  mercury in, 91, 109
coal-fired power plants, 140, 238
  mercury from, 91, 109
cobalt, 114
Colat, Peter, 12
collagen, 166
  N-15 analysis of, 153
Combs, Luke, 74
compasses, 65
computer screens, light from, 135
continents, 179
cooperative systems in nature, 173–76
Cornell University, rock concert study,
    41–42
cosmic rays, 69, 73, 83, 215–19
Cotonou, Benin, 96–97
covalent bonding, 4, 43
Cox, Brian, 7
Crab Nebula, 61, 67–69
cremation, 89, 221–23
  dispersal of atoms during, 221–27
cyanide poisoning, 79
cyanobacteria, 137–38, 177, 185–87
cytochromes, 79, 80

Dalton, John, 40
Darwin, Charles, 188–89
Dawkins, Richard, 213
dead zones in bodies of water, 187, 203
death, 220–32
  annual number of deaths,
    worldwide, 228
  dissipation of atoms in, 214–15
  inevitability of, 206, 227–28
  survival of atoms after, 5–6

decay and decomposition, 148, 149,
    172–73, 221
defensive neurochemicals, 130
deforestation, and carbon dioxide
    emissions, 25
dehydration, 123–26
delta C-13, 105–9
Dennett, Daniel, 213
detergents, phosphate, 185, 193–94
deuterium (H-2), 46–48
  content in living creatures, as
    marker, 47–48, 54
diamonds, 91–92, 225
diatoms, 27
  windborne, from the Sahara,
    188–89
diffusion, 52, 125
digestive enzyme, 52–53
digestive tract, regeneration of cells in,
    211
dinoflagellate algae, 192
DNA, 195
dolphins, 80
Dorneanu, Lucian, 220–21
double helix, 195
doves and cacti in Arizona, 54
dust
  motes in air, 39–40
  windborne, from the Sahara, 188–92

Earth, the
  blue, green, and white colors of,
    from space, 56–57
  climate of, 58
  formation of, 177
  future of, 228–29
  geologic history of, 177–79
  human impact on, 6–7, 179–81
  an ideal planet for humans, 197
  life-sustaining presence of, 241–42
eating and digestion, 149–51
Edison, Thomas, 13, 35

The Effect of Gamma Rays on Man-in-the-Moon Marigolds, 9–10
egg cells (oocytes)
  regeneration of, 212
  size of, 210
Ehleringer, James, 47–48
Einstein, Albert, 3, 19, 40, 62, 133, 136,
    141, 146, 223, 226–27, 230, 233–46
  and the atomic bomb, 240
  family life, 235–36
  as a musician, 242–45
  summer vacations of, in the
    Adirondacks, 234–44
Einstein, Elsa, 235–37
Einstein, Maja, 239
Eisner, Thomas, 116
Electron Microscopy Web site, 8
electrons, 4–5, 134–35
elements
  formation of, in stars, 70–74
  found in human body (you), 5
elephants, 54–55
enamel, dental, 165–66, 212
Energy Information Administration, 99
environmentalism, 7
Environmental Protection Agency
  (EPA), 91
enzymes, 52–53, 80, 138
epidemics, 61
Erickson, George, 142
erosion, 178–79
ethics of land use, 180
eutrophication
  geologic, 201–2
  human-caused, 186, 188, 193
Experimental Lakes Area (ELA),
  184–86
explosives, 132, 142
eyes
  color vision, 136
  lenses of, 212
  moistening of, 111–13, 130–31
  oxygen taken in through, 12–13

Falkowski, Paul, 89
Faraday, Michael, 91–92
fast-food outlets, 105–7
fat cells, regeneration of, 212
Fennelly, Beth Ann, 114–15
Ferguson, Brennan, 174
ferritin, 81–82
ferrous iron, 81
fertilizer
  from Chilean deposits, 142
  nitrogen, 141, 184
  phosphate, 200, 204
  windborne, from Sahara dust,
    190
fetal hemoglobin (hemoglobin F), 79
fingernails, 155
  composition of, 209
Finland, 107
Finney, Bruce, 152
fire, similarity to life, 17–19
fish, mercury in, 91
fluorine, in teeth, 165
food
  elements in, 49, 134
  "junk" and "supermarket," 108
  modern sources of, 156
food chains
  marine, 148–51, 192
  mercury in, 239
  N-15 in, 148–55
Ford, Henry, 13, 35
forests, 22, 25
fossil fuels
  atoms of, in your body, 86–88, 91
  burning of, carbon dioxide
    emissions from, 25, 97–98
  geologic origin of, 100–101
  and global warming, 87
  quantity of, still buried, 101
  running out of, 196
fossils, 56, 100–101, 159, 164–65
four elements (Greek concept), 2
Freehold, N.J., 26

freeloaders and cheating, 175
Freeman, Aaron, 231–32
Freitas, Robert, 197
Frisén, Jonas, 211
Fugate, Martin, 74–76
fugu, 129–30
fungus, 174
future, the, 219, 228–30

gasoline, carbon content of, 16
gas state of matter, 18
Gatorade, 127
genes, immortality in, hoped for, 228
genetic code, 195, 218
geomancy, 64
geophagy (eating of clay), 114–15
Gerlach, Terrance, 98
Germany, 141–46
ghosts. *See* soul or ghost
globalization, 105–8
global warming, fossil fuels and, 87
gold
    atoms, 8
    formation of, in stars, 73
Goran, Morris, 146
granite, quarrying of, 170–71
gravity, and star collapse, 71–73
Great Oxidation Event, 177
Great Recession, 99
Great Rift Valley, 158–59
Greece, ancient, science in, 2–3, 40
Greenblatt, Stephen, 40
green color of algae, 183
Greene, Brian, 7
"green fire" (Aldo Leopold's), 168
greenhouse gases, 157
Greenland, Norse habitation of, 153
Gulf of Mexico, 187, 194
Gulf Stream, 201
gunpowder, 143

Haber, Clara, 144–45
Haber, Fritz, 132–33, 141–47
Haber, Hermann, 145, 146
Haber-Bosch process, 142
    ammonia output of, 155
hair
    carbon-13 in, 107
    composition of, 209
    curly, water in, 44
    proteins of, 53–56
    water in, 44–46, 48–49
Hamblin, Terrence, 77
Haralson, John, 143–44
Hazen, Robert, 176–77
hearing, 129
heart, regeneration of cells in, 210–11
heat, of atoms, 41–42
helium, formation of, in stars, 31, 70
hematite, 66
heme molecules, 77–79
hemlocks, 151
hemoglobin, 28, 63, 75, 77–82
Heussner, Alexandra, 187
Hiroshima bombing, 239
Hitler, Adolf, 146
Holmes, Oliver Wendell, 198
Homo erectus, 160
human body (you)
    atoms composing, 1–2, 5, 9
    after death, 220–27
    elements found in, 5
    like a musical composition, 245–46
    mixed feelings about atomic reality of it, 206
    temperature maintenance of, 18
    temporary nature of, 206–15
human species
    ancestors of, 158–65
    carbon footprint of, 98
    impact of, on Earth, 6–7, 179–81
    Law of the Minimum applied to population of, 192, 196–99, 202–4

as part of nature, 180
population growth made possible by
    nitrogen fertilizers, 147
hydrogen
    atomic structure of, 36
    deep future of, 229–30
    dispersal from body after death,
        222, 224
    formation of, after the Big Bang,
        30, 37
    in the human body, 42–50, 198
hydrogen-2. *See* deuterium
hydrogen-3. *See* tritium
hydrogen bond attraction, 43
hydrogen peroxide, 80–81

Ibn Butlan, 61, 73
ice ages, 202
ice cores, ancient air in, 69
immortality, 228, 230–32
Industrial Revolution, 25, 97, 105
inflammation response, 82–83
inland seas, drying of, 119–21
insecticides, 130
insects, 130
International Fertilizer Development
    Center, 204
iron, 62–85
    abundance of, in the universe, 63
    atomic structure of, 63
    destructive nature of, 62–63
    dispersal from body after death, 227
    ductility of, 63
    early use of, in Iron Age, 83–84
    electric conduction of, 64
    formation of, in stars, 70–74
    in the human body, 66, 76–82, 192
    magnetic, 84
    from plants, 77, 83, 169
    from space, 66–67
    wide use of, in modern world, 84–85
    windborne, from the Sahara, 190–92

iron supplements, 76, 81
ironworking, 64
isoscapes, 47
isotopes
    radioactive, dating by, 240–41
    stable but unusual, 46–47
    tracing by, 148–55, 210

Japan, 107
Jasechko, Scott, 57
Jews, German, 141, 146
junk food, wildlife feeding on, 108

Kaku, Michio, 7
Kansas, 156
Kaopectate, 115
Keeling, Charles David, 23
Keeling, Ralph, 22–25, 27, 32, 87, 99
Kentucky, blue people of, 74–76
keratin, 53–56, 140
Kestenbaum, David, 213
Kheir, John, 12
Kiers, Toby, 175
Kimeu, Kamoya, 160–61
kinesin, 52
King, Gilbert, 144
kit fox, 107–8
Knollwood, in the Adirondacks, 237

lactoferrin, 82
Lagrange, Joseph-Louis, 16
Lake Chad, 189
Lake Turkana, Kenya, 158–59
land, erosion of, 178–79
Large Hadron Collider, 3
Lavoisier, Antoine, 15–16
Law of the Minimum, 183–85
    and human population, 192, 196–99
Leakey, Richard, 158
legumes, 138

Leopold, Aldo, 168–69, 180
Le Rossignol, Robert, 141
Lewis, Richard, 239
Liebig, Justus von, 183–84
life
    analogy with a river, 207
    analogy with combustion, 17–19
    death (nonlife) is more normal, 228
    definition of, 244
    end of, in the universe, 228–29
    origin of, on earth, 242–44
    recycling of elements by, 32
    study of, 7–8
light, effect on human physiology, 135
lightning, 138–39
lightning rod, 64
limestone, 178
Lindemann, Bernd, 211
liquid state of matter, 18
lithium, 122
Little Ice Age, 153
liver (organ), 80
liver cells, 210
lodestone, 64
Los Angeles, California, 95
Lower Saranac Lake, in the
    Adirondacks, Einstein's summer
    retreat at, 234–44
Lucretius, 40
lungs, 19, 53
    functioning of, 17
Lyme disease, 83
Lynnerup, Niels, 153
lysozyme, 127

Magnes, Frances, 243
magnesium, 120
magnetic field, Earth's, 83
magnetism, 64–66, 84
magnetite, 64, 66
Mailer, Norman, 101
manganese, in bacteria, 83

manure, 140
marine mammals, 80
Mars, temperature on, 58
Marshall, Robert, 237
Martinelli, Luiz, 105–7
mastodons, 55–56
matter
    emptiness of, 4
    three states of, 18
Mauna Loa Observatory, Hawaii, 23
McCance, Robert, 122–23
McDonald's restaurants, 102–4
McDonough, Patty, 44
McGee, W. J., 124–25
meat
    color of, when cooked, 79–80
    iron in, 83
melanin, 78
melatonin, 135
mercury
    from coal-fired power plants,
        91, 109
    in dental fillings, 222
    dispersal of atoms from fillings,
        after death, 222
    in fish and humans, 91
    in the food chain, 239
metabolism, process of, 17–19
meteorites
    hunters and collectors of, 66
    minerals and dust from, deposited
        on Earth, 66–67
    Russian, of February 2013, 66
Michelangelo, 243–44
microbes, nitrogen-fixing by, 137–38
Microcystis, 187
microscopes, electron, 38
Miller, Arthur (historian), 243
minerals
    of biological origin, 176–79
    number of kinds of, 177
mist, on a cold morning, 44
Mitchell, Joni, 6

mitochondria, 21, 196
  operation of, 53, 79
Mohr, Hans, 30
molecules, 5
  origin of word, 38
Mongolian spot, 76
Mono Lake, California, 177
Moon, the, temperature on, 58
Morocco, phosphate resources, 202, 204
Morton Salt Company, 118, 121
mothers
  nursing by infant, 154–55
  See also pregnancy
moths, 110–12, 116–17
Mount Saint Helens, Washington State, 98
mouth-to-mouth rescue breathing, 20
Mozart, W. A., 243–45
mummies, 49
muons, 216–17
muscles, 50, 80
  regeneration of cells in, 212
mushrooms, 174
music, 242–45
mutations, 218, 228
mycelium, 174
mycorrhizae, 173
myoglobin, 80

Nagasaki bombing, 239
NASA space program, 31–32
National Oceanic and Atmospheric
  Administration, 98
natural gas, geologic origin of, 101
nature, and the modern world, 103
Naziism, 146
Nelson, Erle, 153
neodymium, 65–66
nerve cells
  Brownian motion in, 52
  regeneration of, 211

nerve impulses, 128
nervous system, 127–30
  regeneration of cells in, 211
  sodium's role in, 127–30
neurons, 128
neutrinos, 216
neutrons, 4
Newton, Isaac, 3
New York City, 26
nickel, formation of, in stars, 72
nicotine, 130
nitrogen, 132–57
  abundance of, in air, 133
  atomic structure of, 134, 143
  conversion to carbon-14, 210, 217
  discovery of, 143
  dispersal from body after death, 222, 224
  fixing of, 137–47
  in food, 134
  formation of, in stars, 71
  in the human body, 137
  oceanic, brought to land by salmon, 149–55
  from plants, 169
nitrogen-15, 148–55
nitrogenase, 138, 190–91
nitrogen compounds, in nature, 138–40, 192
nitrogen cycle, global, 156–57
nitrogen fertilizer, 141, 184
nitrogen molecule (dinitrogen)
  breathing of, 133–34
  instability of, 33
nitrogen oxide
  produced by cosmic rays, 69, 73
  produced by lightning, 138–39
nitrous oxide, in the air, 140–41, 157
Nobel, Alfred, 132
Norse, 153
Nowak, David, 26
nuclear attraction, 71–72

nuclear fusion, in stars, 30–31, 70–73,
    230
nucleotides, 195

oceans
    carbon dioxide in, 89
    nitrogen in, 149–52
oil, geologic origin of, 101
orchids, 175
organic matter
    ancient beliefs about energy of, 37
    decay of, 172–73
Orion Nebula, 69–70
osmosis, 52, 125
oxygen, 11–35
    abundance of, in atmosphere, 25–26
    amount of, in human body, 13–14
    atomic structure of, 3–4
    in blood, 19, 78–80
    in breathing process, 11–13, 19–22
    component of atmosphere, 19,
        23–26, 32
    daily consumption requirement, 11
    discovery of, 14–16
    dispersal from body after death, 224
    in early Earth, 177
    emergency injection of, 12
    formation of, in stars, 31, 71
    in the human body, 13–14, 19, 78–80,
        198
    isotopes of, 55–56
    produced by plants, algae, and
        cyanobacteria, 22, 26–29, 169
    pure, inhaling of, 12
    in rocks, 171
    sources of, 26–30
    toxicity of, 14
oxygen bars in air-polluted cities, 17
oxygen dispensers for home use, 17
oxygen gas, 5
    instability of, 33
    worldwide production of, 29–30

Pankenier, David, 60
Parks, Catherine, 174
parrots, 114
Pataki, Diane, 94–95
Paul Smith's College, 66, 92–93
PCBs, in food chains, 151
Pee Dee, S. C., belemnites, 105
pellicle, 165
pepper (jalapeño), 130
perfume in the air, 52
permethrin, 130
pesticides, in food chains, 151
petroleum-based products, 101–2
phlogiston, 15
phosphate, recycled from human
        waste, 45–46, 204–5
phosphate detergents, 185, 193–94
phosphate fertilizer, 200, 204
phospholipid membrane molecule, 194
phosphorus, 182–205
    abundance of, in the universe, 184
    atomic structure of, 193
    in ATP, 195–96
    in cell membranes, 193–95
    dispersal from body after death, 227
    formation of, in stars, 71, 73
    geologic sources of, 199–204
    name of element, 183
    plankton's need for, 184–87
    from plants, 169, 175
    shortage of, limiting human
        population growth, 192, 199, 202–4
    in the skeleton, 162–63, 167, 171, 192
photosynthesis, 23, 29, 169
Pinatubo volcano, Philippines, 98
Pittman, Rickey, 143
plagues, 82
plankton
    airborne nutrients of, 190–92
    as dietary supplement, 182–83
    fossilization of, 101
    need for phosphorus, 184–87
    oxygen produced by, 27

plants
  carbon compounds produced by, 93
  cooperation with fungi ("plant-
    fungal marketplaces"), 173–76
  elements in, 169, 173–75
  fossilization of, 100–101
  Law of the Minimum applied to,
    183–85
  oxygen produced by, 22, 26–29, 169
  urban, 96–97
  water evaporation from, 57
plasma state of matter, 18
plastic, 102
poison gas, chlorine, 121, 144–46
pollution, air, 17, 32
Popeye (cartoon character), 77
Portugal, 107
Potash Corporation of Saskatchewan
  (PotashCorp), 200
potassium
  dispersal from body after death,
    223
  formation of, in stars, 71
  in the nervous system, 129
  source of, in plants, 114
potassium-40, 219
potatoes, 115
Poulin, Lewis, 140
precipitation, deuterium in, 47
pregnancy
  mother's blood shared with fetus
    during, 79
  nutritional cost to mother, 56, 115
  See also mothers
Priestley, Joseph, 15
proteins
  amount of, in human body, 214
  chemistry of, 137
  of hair, 53–56
  replacement of atoms in, 214
protons, 4, 71–72
Proxima Centauri, 69
puffins, 80

pungent umami taste, 113
pyrethrin, 130

quarrying, 170–71
Queen Charlotte Islands, B.C., 149

radiation, from space, 217–19
radioactive dating, 240–41
radioactive decay, 8
"The Radioactive Orchestra," 8
radioactivity
  in rocks, 219
  in water, 218–19
Ramesh, Nepal, 216
rare earth elements, 65
  formation of, in stars, 73
recycling
  of elements, by life, 32
  of human waste, 45–46, 204–5
red blood cells, 51, 82, 126
  regeneration of, 211–12
  size of, 210
red tides, 184, 192
Reimchen, Tom, 148
Ritchie, Patrick and Rachel, 75–76
rocks
  degradation of, 118–20, 169–71
  salt from, 118–20
Roosevelt, Franklin D., 240
Rutherford, Ernest, 3, 239

Sagan, Carl, 7
saliva, 165
salmon
  migration and spawning of, 147–53
  population fluctuations of, 152–53
salt
  abundance of, in the oceans, 120
  in cells, 125–26
  commercial sources of, 118–21

salt (*continued*)

    consumption of, and health, 113

    deficiency, 122–23

    in history, 113

    from plants, 169

Salt Lake City, Utah, 94–96

salt licks, 114

saltpeter, 142, 143

salt taste, 113, 121–22

salt water, 118–20

A Sand County Almanac, 168–69

saxitoxin, 192

Scheele, Carl, 15

Schindler, David, 157, 184–86

Schopfer, Peter, 30

Schutz, Yves, 214

science

    ancient, 2–3, 15, 40, 60–61,
        64–65

    emotion and meaning derived from,
        231–32

    importance of, to human survival,
        2–3, 6–7, 146

Scotland, 106–7

Scripps Institution of Oceanography,
    La Jolla, California, 22–23, 25

seabirds, 80

seals, 80

seasons, and oxygen-carbon dioxide
    concentrations, 23–24, 25

seawater, 120

seaweeds, 27, 148

selenium, 114

Shapley, Harlow, 33–35

Sherman, Laura, 109

Shertz, Stephen, 32

siderocalin, 82

siderophores, 82

sight, 129

silicon, formation of, in stars, 71

silver, in the human body, 198

Silver Springs, New York, salt deposits
    in, 118–21

skeleton

    elements in, 162–63, 167, 171, 192

    function of, 50, 162

    growth and maintenance of,
        166–67

    individualized shaping of, from life
        experiences, 167–68

    regeneration of cells in, 212

skin

    composition of, 209

    flaking of, 195

    regeneration of cells of, 211

skin color

    blue, 74–76

    brown, 78

sky

    ancient beliefs about, 133

    blue color of, 133–37

$SLCO_2$ Web site, 96

sleep, 135

Smedley, Scott, 116

Smith, Elizabeth, 74

Smith, Lesley, 33

SN1006, 69

SN1054 (origin of the Crab Nebula),
    67–70

Snyder, Shane, 46

sodium, 113–31

    atomic structure of, 113–14

    dispersal from body after death, 223

    formation of, in stars, 71

    in the human body, 122–31

    ingestion of, 121–22

    sources of, in nature, 114–21

sodium nitrate, deposits of, 142

soil

    alive nature of, 172–73

    plant-fungal marketplace in, 173–76

Sokolsky, Pierre, 215

solar energy, stored, released in a fire, 18

solar wind, 83

solid state of matter, 18

Song dynasty (China), 60–61, 64–65